5 FACTS A DAY
SCIENCE

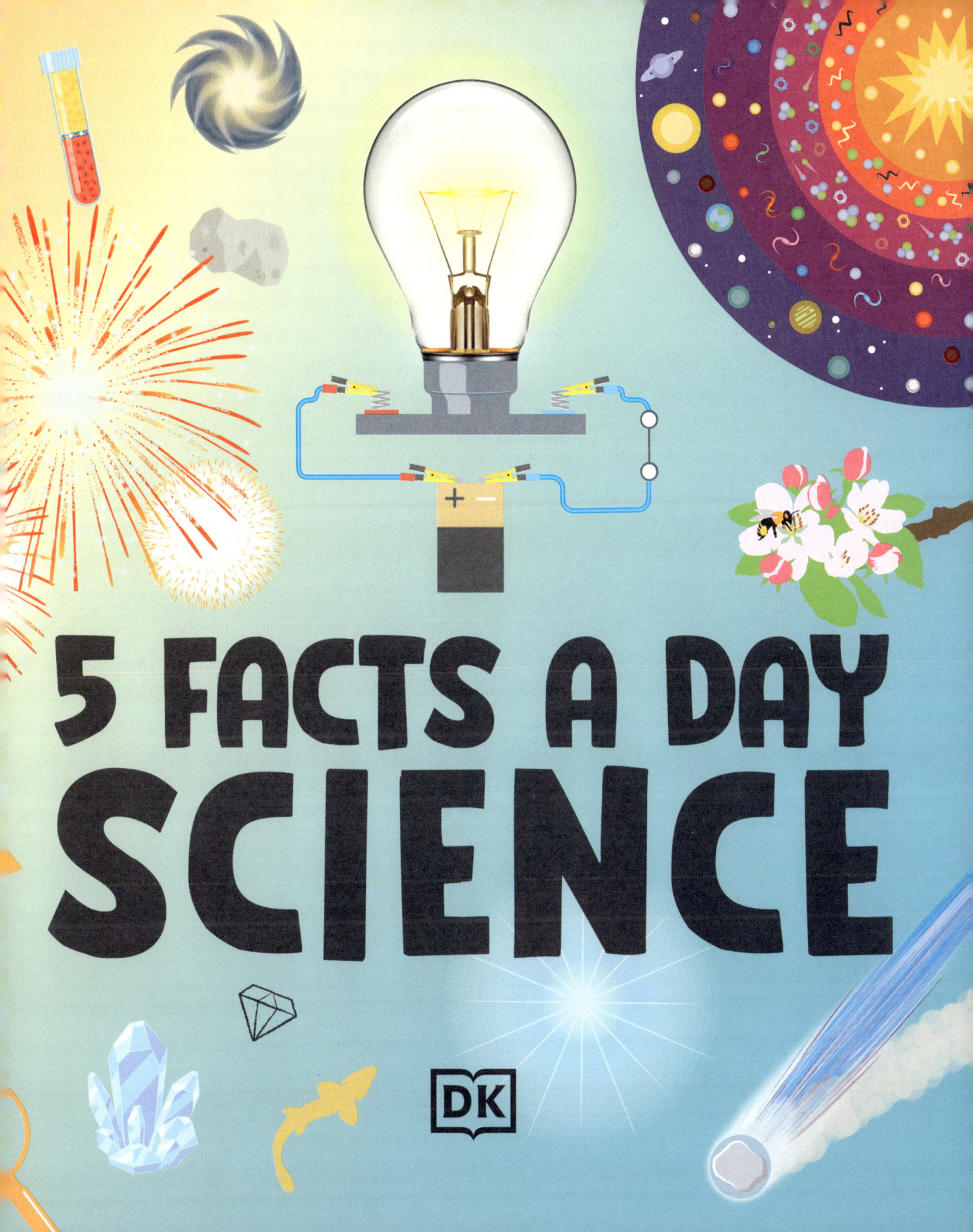

5 FACTS A DAY
SCIENCE

DK

DK | Penguin Random House

Senior Editors Sarah Bailey, Rebecca Fry
Senior Designer Phil Gamble
Additional design support Amy Child
Additional editorial support Vicky Richards, Jenny Sich
Picture Researcher Shubhdeep Kaur
Lead Picture Research Sumedha Chopra
Managing Editor Francesca Baines
Managing Art Editor Philip Letsu
Production Editor Becky Fallowfield
Production Controller Inderjit Bhullar
Jacket Designer Akiko Kato
Senior Jacket Designer Rashika Kachroo
Managing Art Editor (Jackets) Romi Chakraborty
Publisher Andrew Macintyre
Art Director Mabel Chan

Written by Sarah Bailey, Anna Bonnerjea, Rebecca Fry, Francesca Harper,
Ben Morgan, Lizzie Munsey, Vicky Richards, Jenny Sich, Rona Skene

Consultants Josh Barker, Nick Crumpton, Penny Johnson,
Anthea Lacchia, Kristina Routh

First published in Great Britain in 2025 by
Dorling Kindersley Limited
DK, 20 Vauxhall Bridge Road, London SW1V 2SA

The authorised representative in the EEA is
Dorling Kindersley Verlag GmbH. Arnulfstr. 124,
80636 Munich, Germany

Copyright © 2025 Dorling Kindersley Limited
A Penguin Random House Company
10 9 8 7 6 5 4 3 2 1
001–341866–Mar/2025

A CIP catalogue record for this book
is available from the British Library.
ISBN: 978-0-2416-8318-7

Printed and bound in China

www.dk.com

MIX
Paper | Supporting
responsible forestry
FSC™ C018179

This book was made with Forest
Stewardship Council™ certified
paper – one small step in DK's
commitment to a sustainable future.
Learn more at www.dk.com/uk/
information/sustainability

CONTENTS

Emission nebulas emit their own light.

1

A NEBULA is a COSMIC CLOUD OF GAS AND DUST found in the SPACE BETWEEN STARS.

It is in constant motion, mixing and churning material together.

Reflection nebulas don't emit their own light, but reflect the light of nearby stars.

2

NEW STARS ARE BORN inside nebulas called DIFFUSE NEBULAS.

There are three types of diffuse nebula: emission nebulas, reflection nebulas, and dark nebulas.

3

The gas and dust SPEWED OUT BY DYING STARS can form nebulas called PLANETARY NEBULAS AND SUPERNOVA REMNANTS (see p24).

Dark nebulas are so dense they block out light from the stars behind them.

4

The TARANTULA NEBULA spans 1,800 LIGHT YEARS.

That means it would take 1,800 years for light to travel across it!

5

A diffuse nebula THE SIZE OF EARTH would WEIGH JUST 1kg (2.2 lb oz).

This is because there is so much space between the material in a nebula.

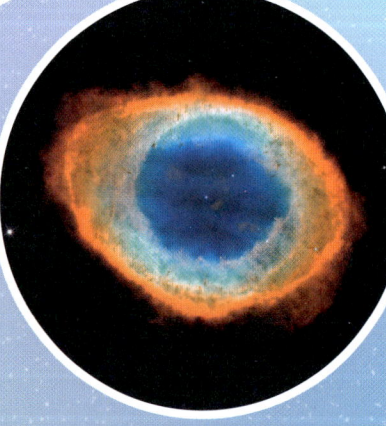

This planetary nebula, formed by a dying star, is called the Ring Nebula.

2
January
ANIMAL WINGS

1 The wings of a WANDERING ALBATROSS are THE BIGGEST OF ANY ANIMAL.

The seabird has a wingspan that measures 3.5 m (11.5 ft) from tip to tip.

2 DRAGONFLIES can move each of THEIR FOUR WINGS INDEPENDENTLY.

This helps them make acrobatic maneouvres like flying backwards as they hunt other flying insects.

3 Many BUTTERFLIES have FAKE EYES on their wings.

They startle predators when the wings open, giving the butterfly a brief chance to escape.

4 SWIFTS' wings are SO GOOD AT GLIDING that the birds can STAY AIRBORNE FOR 10 MONTHS without landing.

They sleep in mid-air and don't land until they are old enough to nest.

5 PENGUINS flap their wings to FLY THROUGH THE WATER.

They are the world's fastest swimming birds, capable of 35 km/h (22 mph) underwater.

3
January
BUOYANCY

1 TWO FORCES act on an object in water – its own WEIGHT pulling it down and the force of UPTHRUST pushing it up.

WEIGHT

UPTHRUST

2 When an object is submerged in water, it DISPLACES (pushes aside) an amount of water EQUAL TO ITS OWN VOLUME.

Displacement results in an upwards force from the water called upthrust.

3 Ancient Greek mathematician ARCHIMEDES was the FIRST TO DISCOVER DISPLACEMENT.

He noticed his body displaced the water when getting into the bath, and is said to have yelled "Eureka!".

4 If the MASS OF WATER DISPLACED IS MORE THAN the MASS OF THE OBJECT, the upthrust will be greater than the weight, and THE OBJECT WILL FLOAT.

WEIGHT

A heavy weight is denser than water, so it sinks.

UPTHRUST

5 Objects that are LESS DENSE THAN WATER will FLOAT ON THE SURFACE, while DENSER OBJECTS WILL SINK.

4 January EARLY LIFE

1 LIFE may have EMERGED from DEEP IN THE OCEAN.

Some scientists think minerals in deep-sea vents (see p130) mixed with ocean water under extreme pressure to produce the building blocks of life.

2 ALL LIFE on Earth probably evolved from a SINGLE MICROBE, known as LUCA, 4 billion years ago.

3 LUCA led to the two main groups of life: BACTERIA and ARCHAEA.

These life forms continued to evolve into others over billions of years.

Deep-sea hydrothermal vent (see p130)

Chemicals in the water react with minerals in the vent, creating the building blocks of life.

Hot, chemical-rich water rises through the vent from inside Earth.

4 ANIMALS evolved about 800 MILLION YEARS AGO. The earliest animals were SPONGES.

5 The oldest FOSSILS are 3.7 billion years old.

These microorganisms may be some of Earth's oldest life forms.

5 January NERVES

1 MESSAGES travel around your brain and body via ELECTRICAL SIGNALS carried by long, wiry cells called NEURONS.

2 Signals can travel at up to 400 km/h (250 mph) - AS FAST AS the FASTEST RACING CAR!

3 If you stroke a cat, the SIGNAL telling you it's SOFT reaches your brain 10-20 MILLISECONDS FASTER than the one telling you it's WARM.

Cell body

Nucleus

Axon

Synapse

Axon terminal

Signals travel along the axon as electrical impulses.

The myelin sheath insulates the axon, so impulses travel along it faster.

4 A single neuron can send out as many as 1,000 MESSAGES PER SECOND.

5 All neurons carry signals in just ONE DIRECTION.

Some carry signals from brain to body, others from body to brain.

1 THE TAIGA is the world's LARGEST LAND BIOME.

Also called boreal forest, it is a type of forest that grows across huge areas of northern Europe, Asia, and North America.

6

January
TAIGA

2 The taiga contains a QUARTER of ALL EARTH'S TREES.

It is packed with evergreen coniferous trees, such as fir, spruce, and larch.

3 CONIFER TREES are well adapted for the COLD, DRY taiga climate.

Their needle-shaped leaves can withstand freezing winds and heavy snow, and their waxy coating prevents them from losing moisture.

In winter, most taiga birds, including the snow goose, migrate south to warmer habitats.

4 Taiga WINTERS are long... EIGHT MONTHS LONG!

Some animals, such as the grizzly bear, hibernate through the winter when there is very little food available.

Thick fur and layers of fat protect brown bears from the cold.

A conical shape and downward-sloping branches help snow slide off conifer trees.

Moose have thick, insulating fur, and their large hooves prevent them from sinking into the snow.

Keen senses and sharp teeth help wolves hunt large mammals such as moose.

5 The taiga's LARGEST ANIMAL is the MOOSE.

This species of deer has extra-long legs that help it move through the deepest snow and wade in forest lakes.

7
January
YOUR HEART

1 The heart is a fist-sized MUSCULAR ORGAN that PUMPS 5 l (9 pt) of BLOOD around the body every minute.

In one year, it pumps enough blood to fill an Olympic swimming pool.

2 Your heart works tirelessly, beating around 100,000 TIMES A DAY – that's 2.5 BILLION BEATS by the time you are 70.

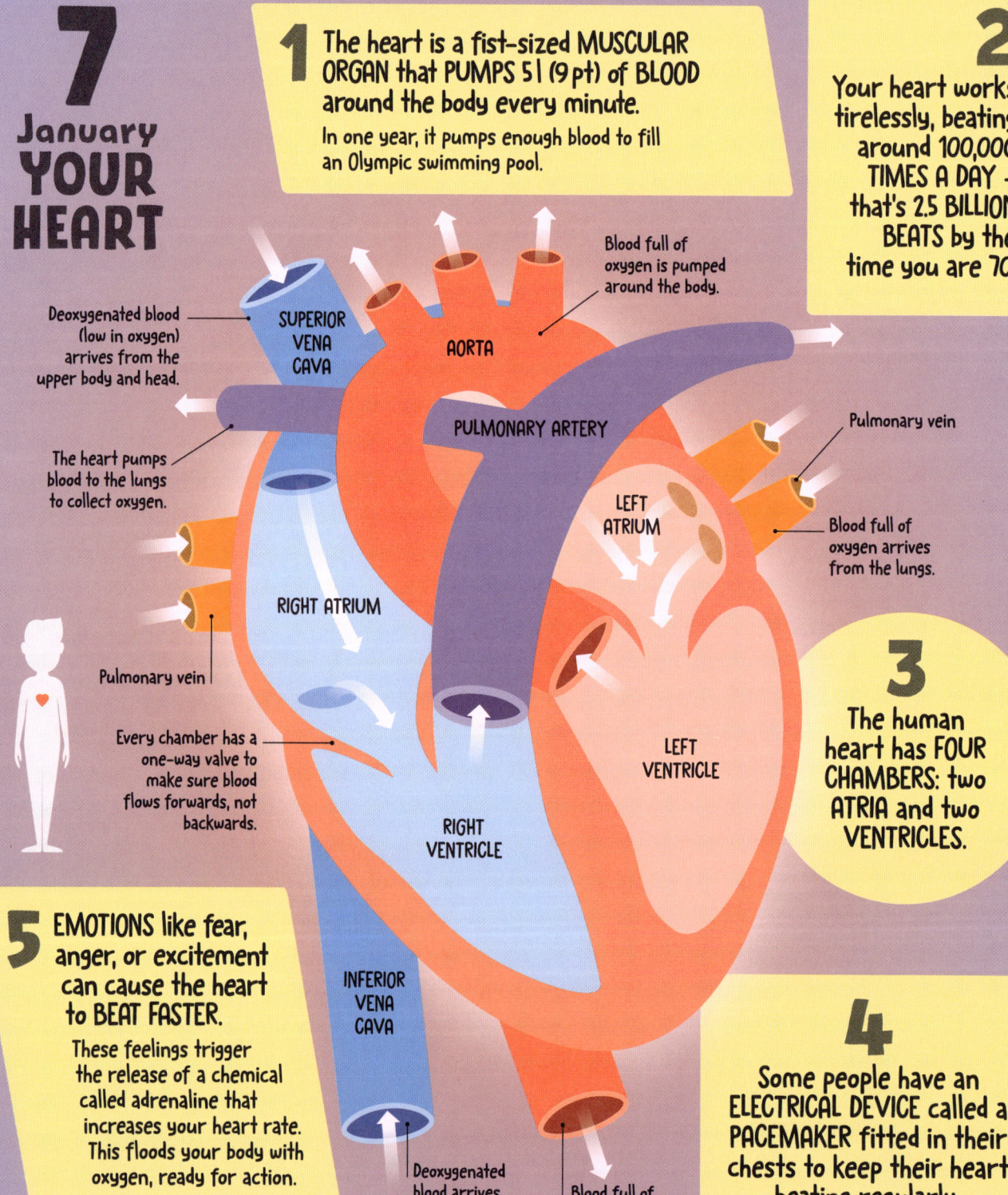

Deoxygenated blood (low in oxygen) arrives from the upper body and head.

The heart pumps blood to the lungs to collect oxygen.

Pulmonary vein

Every chamber has a one-way valve to make sure blood flows forwards, not backwards.

SUPERIOR VENA CAVA

AORTA

Blood full of oxygen is pumped around the body.

PULMONARY ARTERY

RIGHT ATRIUM

LEFT ATRIUM

Pulmonary vein

Blood full of oxygen arrives from the lungs.

LEFT VENTRICLE

RIGHT VENTRICLE

INFERIOR VENA CAVA

Deoxygenated blood arrives from the lower body.

Blood full of oxygen goes to the lower body.

3 The human heart has FOUR CHAMBERS: two ATRIA and two VENTRICLES.

5 EMOTIONS like fear, anger, or excitement can cause the heart to BEAT FASTER.

These feelings trigger the release of a chemical called adrenaline that increases your heart rate. This floods your body with oxygen, ready for action.

4 Some people have an ELECTRICAL DEVICE called a PACEMAKER fitted in their chests to keep their heart beating regularly.

8
January
VENOM

Venom flows out through a hole in the tip of each fang.

1 Venom is a TOXIC CHEMICAL injected through a BITE or STING.

It is different from poison, which is swallowed, inhaled, or absorbed through the skin.

A snake's venom gland produces venom.

2 Most animals use venom to PARALYSE and KILL PREY or FEND OFF hungry PREDATORS.

3 Some venoms can be USED IN MEDICINE.

The excruciating venom of the gila monster contains a chemical used to treat diabetes in humans.

4 There are 725 SPECIES of venomous SNAKES.

They use sharp, hollow fangs to deliver their venom.

5 JELLYFISH INJECT prey with venom from special cells called NEMATOCYSTS.

They consist of a venom-filled capsule and harpoon-like thread that pierces their prey.

Nematocysts line the sea nettle's long tentacles.

9
January
TREES

2 NOT all trees LOSE THEIR LEAVES in winter.

Evergreens have thinner, needle-shaped leaves that stay on their branches all year round.

Count the rings spanning out from the centre of the trunk.

3 You can tell a tree's AGE by COUNTING the RINGS in its trunk.

The more rings, the older the tree.

Each ring marks one year of a tree's life.

1 The TALLEST TREE in the world, Hyperion, is over 116 m (380 ft) tall – TALLER THAN THE EIFFEL TOWER!

4 Trees in a FOREST may be able to COMMUNICATE with each other.

Their roots are linked by fungi, which can pass signals and nutrients between the trees.

5 Trees can become FOSSILS. Instead of decaying, some dead trees can become PETRIFIED and TURN TO STONE (see p34).

1 A volcano forms when MOLTEN ROCK, called MAGMA, escapes through an opening in Earth's crust and SOLIDIFIES as it cools.

2 A volcano's SHAPE depends on the LAVA that ERUPTS from it.
Thick lava hardens quickly to form cone-shaped mountains. Runny lava spreads and creates flatter mounds.

3 When PRESSURE BUILDS UP below the surface, it can result in violent volcanic ERUPTIONS.

Eruptions can be sudden and explosive.

Lumps of lava cool and harden into rock mid-air.

Gas, steam, and ash erupt from the volcano.

Magma that has erupted through the surface is called lava.

Thick clouds of extremely hot gas, ash, and rock, called pyroclastic flows, burn anything in their way.

Some lava may erupt through secondary vents.

An eruption begins when magma rises through the main vent from a magma chamber below.

4 Many of EARTH'S ISLANDS were made by UNDERSEA VOLCANOES.
All of Hawaii's 137 islands were created by volcanic eruptions on the seafloor.

10 January VOLCANOES

5 VOLCANOES can cause LIGHTNING!
Ash from an eruption causes a build-up of static electricity in the clouds, which is discharged as lightning.

11 January SNAKES

1 A snake's SCALES are HARDENED FOLDS OF SKIN.

A snake grows out of its skin every few months.

The brain interprets information from the Jacobson's organ.

The snake also smells through its nostrils.

Scent molecules are transferred to the Jacobson's organ.

The tongue collects odour molecules from the air.

5 The spider-tailed VIPER has a tail that looks like a SPIDER.

The viper waves its tail to attract spider-eating birds, then strikes.

Pit organ

2 A snake smells with its TONGUE.

It picks up odour molecules from the air and transfers them to a special organ in the mouth that detects smell.

3 The reticulated python is the LONGEST SNAKE IN THE WORLD.

It can grow to 10 m (32 ft) long and can catch and swallow a wild pig.

4 Some snakes can "SEE" HEAT.

The pit organ detects infrared radiation (see p256) to create a thermal image, so the snake sees its prey in the dark.

12 January INSIDE EARTH

Solid crust

Solid upper mantle

Putty-like lower mantle

Liquid outer core

5 The outer core is made of CHURNING, MOLTEN IRON AND NICKEL, which generates a MAGNETIC FIELD.

Solid inner core

1 Earth formed about 4.5 BILLION YEARS AGO.

2 The TEMPERATURE of Earth's inner core is 5,500°C (9,950°F) – as hot as the SURFACE OF THE SUN.

3 The ROCKY SLABS of Earth's CRUST are FUSED to the UPPER MANTLE. The softer lower mantle is constantly moving.

4 If Earth were the size of an APPLE, its CRUST would be the same thickness as the APPLE'S SKIN.

15

1

All animals, including humans, ARE MADE UP OF MILLIONS OF MICROSCOPIC CELLS.

Nobody knows exactly how many, but there could be as many as 30 trillion (million million) inside you!

2

There are lots of different types of cell (see p214), but most HAVE THE SAME STRUCTURES INSIDE them called "ORGANELLES", which DO SPECIFIC JOBS.

The nucleus is the cell's control centre.

3

MOST TYPES OF CELL are SO TINY that 200 of them WOULD FIT INSIDE A FULL STOP.

Mitochondria are the cell's powerhouses. They use oxygen to break down glucose to release energy.

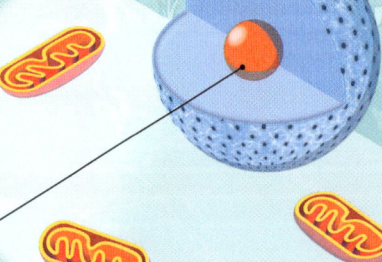

The cell is full of a jelly-like fluid, in which all the organelles float.

Genes (see p29) inside the nucleus contain instructions that tell the cell what to do.

The cell membrane regulates the flow of substances into and out of the cell.

4

An ovum (female egg) is the LARGEST HUMAN CELL.

With a diameter of 0.1 mm (0.04 in), it is just big enough to see without a microscope.

13

January

ANIMAL CELLS

5

BLOOD CELLS DON'T HAVE A NUCLEUS.

This leaves more space inside to carry oxygen.

1 ONE-THIRD of Earth's land is COVERED BY DESERTS.

Every continent has at least one desert, and some have several.

2 A DESERT can be HOT or COLD.

It is any area that receives less than 2.5 cm (1 in) of rain, snow, or sleet per year.

3 About 11,000 years ago, the SAHARA was NOT A DESERT.

It was green and grassy, with lakes, rivers, and even forests. Scientists are trying to work out why it became a desert.

14
January
DESERTS

4 ANTARCTICA IS A POLAR DESERT and the world's LARGEST DESERT.

The Sahara is the world's largest hot, sandy desert.

5 STRONG WINDS in deserts around the world cause SAND and DUST STORMS called HABOOBS.

They can be thousands of kilometres wide, and lift sand and dust up to 1,500 m (5,000 ft) in the air, which gets blown across continents.

15
January
AMAZING CARS

1 THRUST SSC BROKE THE LAND SPEED RECORD IN 1997.

It reached 1,227 km/h (763 mph), broke the sound barrier (speed of sound), and created a sonic boom!

2 Inspired by a James Bond movie, a Swiss company showcased THE WORLD'S FIRST UNDERWATER CAR IN 2008.

Sadly, it was only a concept car and is not available to buy!

4 In 2018, a Tesla Roadster WAS SENT INTO SPACE and placed in orbit around the Sun.

It went up on board the SpaceX Falcon Heavy rocket as a marketing stunt, with a "starman" dummy at the wheel.

3 The smallest car to make it onto roads was the Peel P50 – JUST 137 cm (54 in) LONG. It was manufactured in the 1960s and HAD ROOM FOR JUST ONE PERSON.

5 The FASTEST CARS available to buy today can reach SPEEDS OF MORE THAN 483 km/h (300 mph) and GO FROM 0–97 km/h (0–60 mph) IN JUST 2.5 SECONDS!

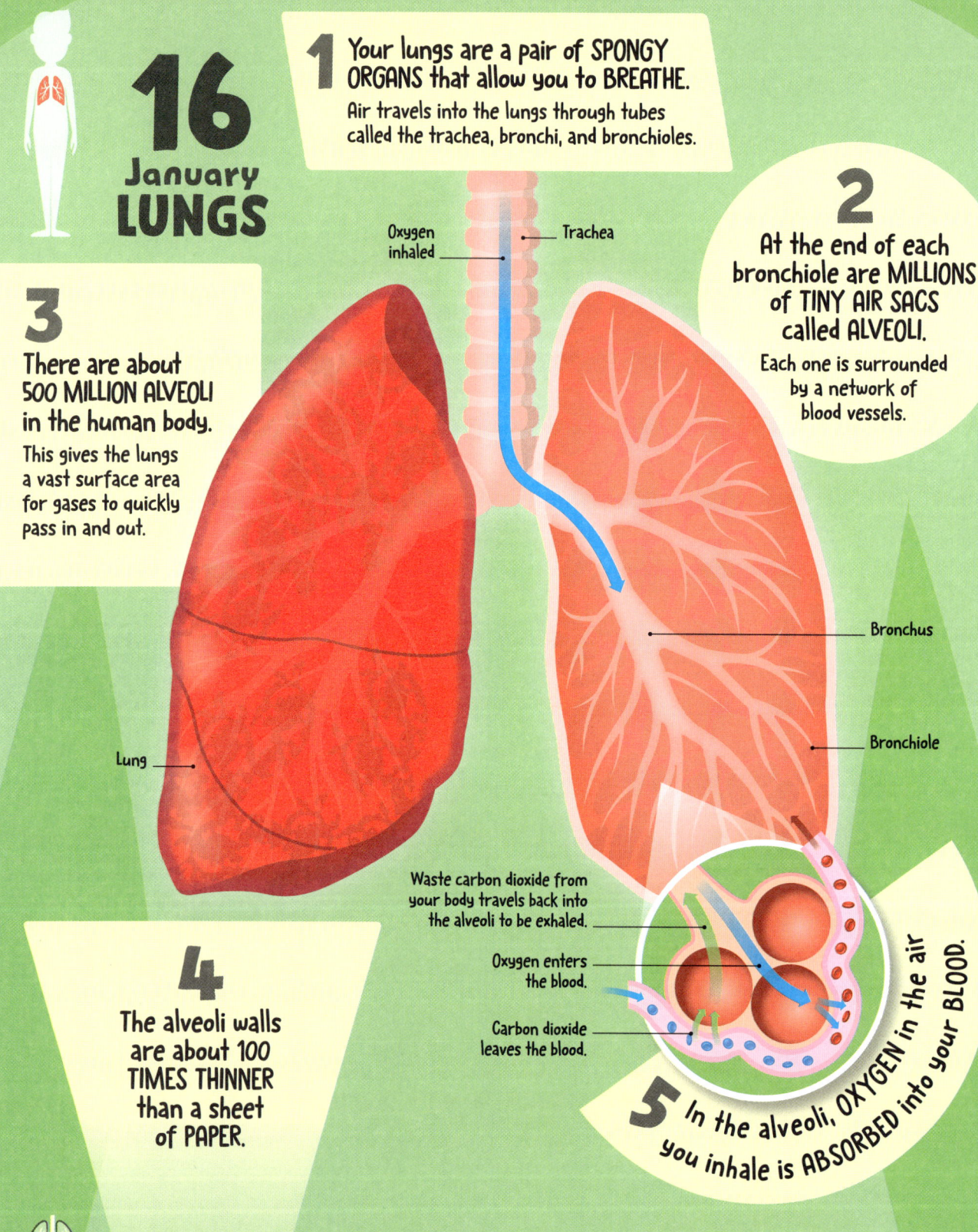

16
January
LUNGS

1 Your lungs are a pair of SPONGY ORGANS that allow you to BREATHE.

Air travels into the lungs through tubes called the trachea, bronchi, and bronchioles.

2 At the end of each bronchiole are MILLIONS of TINY AIR SACS called ALVEOLI.

Each one is surrounded by a network of blood vessels.

3 There are about 500 MILLION ALVEOLI in the human body.

This gives the lungs a vast surface area for gases to quickly pass in and out.

4 The alveoli walls are about 100 TIMES THINNER than a sheet of PAPER.

5 In the alveoli, OXYGEN in the air you inhale is ABSORBED into your BLOOD.

Oxygen inhaled

Trachea

Bronchus

Bronchiole

Lung

Waste carbon dioxide from your body travels back into the alveoli to be exhaled.

Oxygen enters the blood.

Carbon dioxide leaves the blood.

17 January METALS

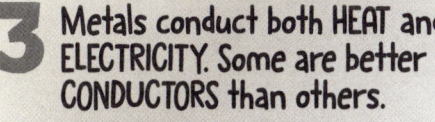

Metal's malleability made it perfect for moulding into suits of armour.

1 Metals are a type of ELEMENT (see pp152–153). They make up more than three-quarters of the periodic table.

2 Most metals share several key properties – they are HARD, SHINY, and can be HAMMERED INTO SHAPE (malleable).

3 Metals conduct both HEAT and ELECTRICITY. Some are better CONDUCTORS than others.

4 If two pieces of metal TOUCH IN SPACE (or a vacuum), they can FUSE TOGETHER! This doesn't happen on Earth because metals exposed to the air have a barrier layer on them.

5 Iron is the MOST COMMON METAL in the Universe. It makes up 11 per cent of matter!

---

1 Our mouths can sense FIVE BASIC TASTES: sweet, sour, salty, bitter, and umami (savoury).

Taste buds are tiny bumps on your tongue that taste food.

2 The average adult has about 10,000 TASTE BUDS.

3 You LOSE taste buds as you AGE. They can also become smaller and less sensitive.

4 SPICY FOODS trick your mouth into feeling like it is BURNING because they contain a chemical called CAPSAICIN.

5 A huge amount of the FLAVOUR we get from food comes from SMELL rather than TASTE.

18 January TASTE

19 January
VIRUSES

1 Viruses are the SMALLEST TYPE OF GERM THAT CAUSE DISEASE.

They are so small that 400 million of some types could fit inside a full stop.

2 VIRUSES ARE NOT REALLY ALIVE.

They are just little packets of genetic information wrapped in a protein shell. They don't eat, grow, or move independently.

3 Viruses survive by HIJACKING A LIVING HOST – human, animal, or plant.

They make the host cells produce millions more copies of themselves.

Genetic information (DNA or RNA) is stored inside the virus.

Surface proteins attach the virus to the host cell.

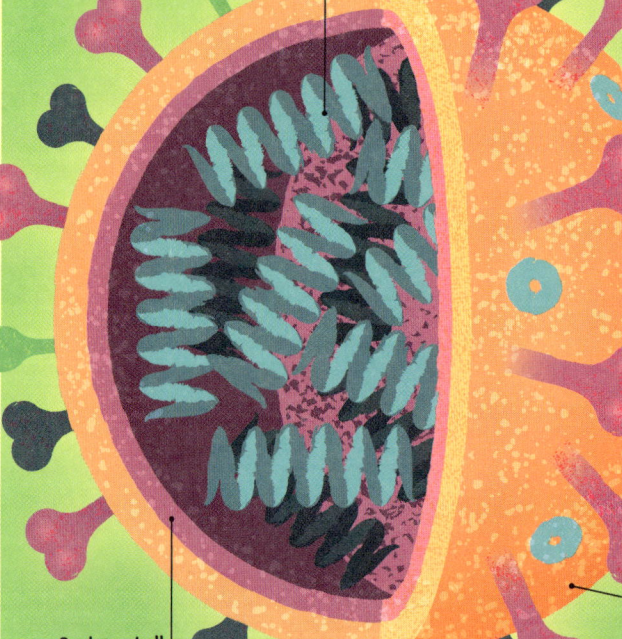

The outer shell protects the virus.

Protein shell surrounds the genetic code.

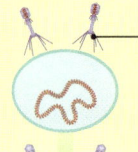
The virus lands on a healthy host cell.

It injects its DNA or RNA into the cell.

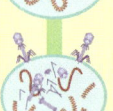
The viral DNA uses the cell to copy its own DNA and proteins.

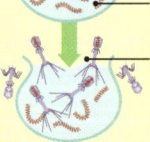
New viruses burst out, killing the cell. They look for more cells to infect.

4 Most viruses MULTIPLY VERY FAST.

This means they can evolve rapidly, for instance to become resistant to antiviral medicines.

5 Viruses come in LOTS OF DIFFERENT SHAPES AND SIZES.

HELICAL

POLYHEDRAL

SPHERICAL

COMPLEX

INHALING (BREATHING IN)

1

Oxygen is a key ingredient in MANY OF YOUR BODY'S ESSENTIAL PROCESSES.

Breathing air into your lungs (see p18) is your body's way of getting the oxygen it needs.

3. Air is pulled into your lungs.

2. Your lungs and chest expand.

1. The diaphragm muscle contracts.

20
January
BREATHING

EXHALING (BREATHING OUT)

6. Air is forced out through your nose or mouth.

4. The diaphragm muscle relaxes.

5. Your chest and lungs reduce in volume.

2

Your lungs and diaphragm ACT LIKE A SUCTION PUMP.

When the diaphragm (a dome-shaped muscle below your lungs) contracts, your chest and lungs expand. This pulls oxygen-rich air into your lungs through your nose and mouth.

3

When you EXHALE, air is FORCED OUT of your lungs.

Your diaphragm relaxes, returning to its dome shape, which in turn reduces the space in the chest and lungs, and forces the air out.

4

On average, AN ADULT BREATHES 22,000 TIMES IN A DAY, if not exercising.

5

Some people SNORE WHEN THEY ARE ASLEEP.

This happens when the soft tissues in their throat and nose relax and vibrate as the person breathes in and out.

1

HUMANS HAVE USED NATURAL PRODUCTS AS MEDICINE FOR THOUSANDS OF YEARS.

From plants and honey, to soil and animal dung, people had to rely on trial and error to discover what worked.

60,000 years ago, people used yarrow plants to treat wounds. It is still used around the world today.

2

PEOPLE once THOUGHT that their PAIN and DISEASE came from EVIL SPIRITS ENTERING THE BODY.

They believed trepanation (cutting holes in the skull) would release the spirits. Trepanation is one of the earliest known surgeries, dating back to 6000 BCE.

People thought trepanation could also treat headaches and epilepsy.

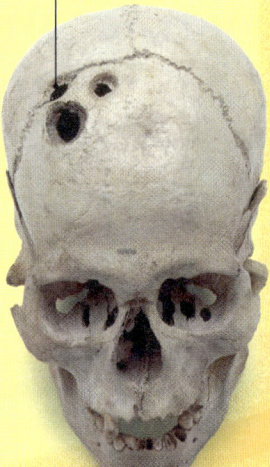

3

The ancient Egyptians KNEW THE IMPORTANCE OF THE HEART around 3500 BCE.

As well as pumping blood, they thought it transported poo, air, and the soul around the body.

21
January
ANCIENT MEDICINE

4

MOULDY BREAD WAS PRESSED ONTO WOUNDS TO TREAT INFECTIONS in ancient Egypt, China, Serbia, Greece, and Rome.

Eventually, in 1928, Alexander Fleming discovered mould could be used to make antibiotics.

5

The ancient Greek Hippocrates believed THE BODY was MADE UP OF FOUR "HUMOURS" (fluids): blood, phlegm, yellow bile, and black bile.

If someone was unwell, he thought it was because their humours were imbalanced.

In medieval Europe, leeches would be used to suck a person's blood to rebalance their humours.

22
January
TIME

1

TIME is a concept that HELPS US TO MAKE SENSE OF AND ORDER EVENTS, and to organize our lives.

2

TIME IS DIVIDED INTO UNITS like seconds, hours, days, and years TO HELP US KEEP TRACK OF IT.

3

IN PHYSICS, TIME IS A DIMENSION LIKE MASS AND LENGTH.

4

According to Einstein's theory of relativity, TIME VARIES DEPENDING ON HOW FAST YOU ARE MOVING.

Time goes more slowly in a fast-moving object.

5

ATOMIC CLOCKS COUNT THE VIBRATIONS OF ATOMS TO MEASURE TIME, which means they are very accurate.

23
January
THE BIG BANG

1 THE BIG BANG IS THE HUGE EVENT THOUGHT TO HAVE CREATED ALL MATTER IN THE UNIVERSE.

2 THE UNIVERSE APPEARED 13.8 BILLION YEARS AGO, beginning as a tiny hot, dense blob that EXPANDED very very quickly.

3 The FIRST PARTICLES probably formed JUST A FEW MILLIONTHS OF A SECOND AFTER THE BIG BANG, but it took another 380,000 years for atoms (see p53) to form.

4 ENERGY FROM THE BIG BANG is still around today. It is known as cosmic microwave background radiation.

5 BIG BANG THEORY WAS PROPOSED IN 1927, but didn't become widely accepted until the 1960s.

24 January EXPLODING STARS

The nebula is made up of gases such as hydrogen, oxygen, and sulfur.

Remnants of a supernova spotted in 1054 CE in the Taurus constellation are still visible today as the Crab Nebula.

1 Some stars END THEIR LIFE AS A SPECTACULAR EXPLOSION called A SUPERNOVA.

2 This happens when A MASSIVE STAR RUNS OUT OF FUEL and DIES.

The star collapses, causing shockwaves so great the star violently explodes.

3 A supernova can trigger the BIRTH OF NEW STARS.

The material thrown out may also become part of the next generation of stars.

4 A supernova can HURL MATERIAL INTO SPACE AT 15,000–40,000 km/s (9,000–25,000 miles/s).

5 A supernova can SET THE SKY ALIGHT FOR WEEKS, but their remnants can remain visible for THOUSANDS of YEARS.

25 January INSECTS

1 INSECTS are ARTHROPODS, which means they have THEIR SKELETONS ON THE OUTSIDE.

Adult insects also have six legs, a three-part body, antennae, and most have two pairs of wings.

2 There are about 1.4 BILLION INSECTS for EVERY PERSON ON EARTH.

3 Most insects have compound EYES, which are MADE OF THOUSANDS OF TINY LENSES.

The many lenses allow the insect to see in a wide angle and detect movement much faster than humans can.

4 Katydids (bush crickets) HAVE EARS ON THEIR FRONT KNEES.

5 Insects have been around ON EARTH FOR MORE THAN 350 MILLION YEARS.

That's longer than the dinosaurs or flowering plants.

Antennae

Head

Thorax

Six legs

Exoskeleton

Abdomen

Two pairs of wings

1

ELECTRICITY IS A WAY OF TRANSFERRING ENERGY.

Humans can generate electricity by using sources such as wind or coal (see p203) to power generators, but it can also occur naturally – shooting down from the sky in giant bolts of lightning (see p193)!

When electricity passes through objects in a circuit, it powers them – causing this bulb to light up.

26
January
ELECTRICITY

2

TO POWER ITEMS such as computers, TVs, and even cars, **ELECTRICITY MUST FLOW IN A LOOP CALLED A CIRCUIT.**

In the circuit, wires connect an object that needs to be powered by electricity to a power source.

The electric current flows one way around a circuit.

5

ELECTRIC EELS can produce electricity to STUN OTHER FISH.

They send shocks of up to 860 volts into the water – enough to knock out their prey, or a human – all while avoiding harm to themselves!

A battery can be the power source of a circuit.

3

TINY PARTICLES called electrons create ELECTRICITY.

These all move in one direction, creating a flow of electricity called electric current.

Minuscule electrons flow between the atoms that make up the wire.

4

ONLY SOME MATERIALS ARE "CONDUCTORS" (electricity flows through them easily).

Wires in a circuit are often copper, which conducts electricity. Plastic and glass are known as "insulators" because they don't conduct well.

For the electricity to flow, the switch must be turned on, connecting all parts together and closing the circuit.

1

THE ATMOSPHERE is one of the main reasons why LIFE ON EARTH is possible.

It traps warmth from the Sun and protects us from harmful radiation and objects from space.

The exosphere extends to about 10,000 km (6,200 miles) from Earth's surface.

EXOSPHERE

GPS (see p95) and other medium-orbit satellites orbit here.

27
January
THE
ATMOSPHERE

2

The atmosphere consists of a MIXTURE OF GASES that are HELD AROUND EARTH BY ITS GRAVITY.

They include nitrogen, which is essential for all life, and oxygen, which animals need to breathe.

3

THE ATMOSPHERE IS MADE UP OF FIVE LAYERS.

From the ground up, these are: the troposphere, stratosphere, mesosphere, thermosphere, and exosphere.

Auroras are seen in this layer, over the polar regions.

THERMOSPHERE

The Kármán line marks the official boundary where "space" begins.

As starlight passes through the atmosphere, the rays are bent, making their brightness seem to change.

This is the layer where meteors burn up on entry.

MESOSPHERE

4

THE OZONE LAYER in the stratosphere PROTECTS US from THE SUN'S ULTRAVIOLET RADIATION.

Chemicals called CFCs from things like fridges and aerosol sprays damage the ozone layer, causing "holes". Their use has now been banned.

5

IT'S THE ATMOSPHERE THAT MAKES STARS "TWINKLE".

Ozone layer

Weather balloons gather information here.

All of our weather happens in the troposphere.

STRATOSPHERE

TROPOSPHERE

ACIDS AND BASES

1 ACIDS AND BASES ARE CHEMICAL OPPOSITES. When put together, these different substances react to "neutralize" each other.

2 STRONG ACIDS CAN BURN!

Acids such as hydrochloric and sulphuric acid will burn the skin. Weak acids, such as vinegar and lemon juice, are often found in food, and won't burn.

3 The pH scale measures ACIDITY or ALKALINITY (basicity) of substances.

To find the pH of a liquid, a special indicator can be dipped into it and the colour it turns matched with the pH scale.

4 Common CLEANING PRODUCTS ARE often VERY ALKALINE.

Bleach, oven cleaner, and other household cleaners can irritate your skin because they are so strong.

5 OUR STOMACH CONTAINS A VERY STRONG ACID (with a pH of 2). It breaks down our food and kills harmful bacteria.

PH SCALE

ACIDS
- 0 BATTERY ACID
- 1 STOMACH ACID
- 2 LEMON JUICE
- 3 ORANGE JUICE
- 4 TOMATO JUICE
- 5 BLACK COFFEE

NEUTRAL
- 6 COW'S MILK
- 7 PURE WATER

ALKALIS (BASES)
- 8 SEAWATER
- 9 TOOTHPASTE
- 10 ANTACID TABLET
- 11 AMMONIA
- 12 SOAPY WATER
- 13 OVEN CLEANER
- 14 DRAIN CLEANER

Car battery acid is extremely acidic, measuring about 0.8 on the pH scale.

Lemon juice is around 2.5 on the scale because it contains citric acid.

Milk has a pH of around 6.5. It is slightly acidic due to the lactic acid it contains.

The pH of pure water is 7, which is neutral, so neither alkaline nor acidic.

Toothpaste is alkaline so that it neutralizes acids produced by bacteria in the mouth after eating.

The pH of detergents can vary a lot, but is generally alkaline to remove unwanted grease.

Drain cleaners contain sodium hydroxide, which is extremely alkaline – around 14 on the pH scale.

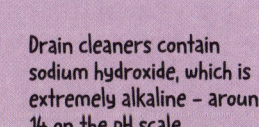

SPIRAL GALAXIES have a dense bulge of old stars at their centre, with arms of newer stars rotating around them.

BARRED SPIRAL GALAXIES, like our galaxy, the Milky Way, have an elongated bulge at their centre.

1 Galaxies are made up of STARS, PLANETS, and CLOUDS of gas and dust particles held together by GRAVITY.

Thousands of galaxies can be held together by their gravity, forming "groups" and "clusters".

2 The largest galaxies can contain TRILLIONS OF STARS!

The smallest galaxies can be made up of just a few thousand stars.

3 There may be as many as 200 TRILLION (two thousand billion) GALAXIES IN THE UNIVERSE!

They can be organized into five main types, based on their different shapes: spiral, barred spiral, lenticular, elliptical, and irregular.

4 THE CLOSEST GALAXY TO OURS IS THE ANDROMEDA GALAXY.

It lies 2.5 million light years away now, but in about five billion years it will collide with us!

LENTICULAR GALAXIES are flat discs with a bulge at the centre, but no spiral arms.

ELLIPTICAL GALAXIES are oval in shape and have little structure.

IRREGULAR GALAXIES are the least common type of galaxy.

5 If we LOOK CLOSELY at even just a tiny sliver of sky, we can see THOUSANDS of galaxies.

This James Webb Space Telescope image shows the galaxies that can be seen in an area the size of a grain of sand held at arm's length!

30 January DNA

DNA is stored in your body's cells.

A DNA molecule is shaped like a spiral ladder known as a double helix.

Chromosome

The "rungs" of the ladder are made of pairs of chemicals called bases.

1 DNA is a CHEMICAL CODE found in all living things.
It is stored in the nucleus of almost every cell in your body, except for red blood cells.

2 DNA molecules are COILED into structures called CHROMOSOMES.
Humans typically have 23 pairs of chromosomes in each cell. Each pair is made up of one chromosome inherited from your mother and one from your father.

3 Over 90 per cent of your DNA is "JUNK" and not used for genes (see below).
Scientists don't yet understand why so much of it seems to have no purpose!

4 EVERYONE'S DNA IS UNIQUE.
Even identical twins have minor differences! This means DNA can help police solve crimes and identify a suspect by analysing DNA from hair, blood, or skin cells left at a crime scene.

5 Your DNA is very TIGHTLY PACKED into cells.
If you unravelled it, the DNA in one cell would measure 2 m (6.6 ft). All the DNA in your body would reach to the Sun and back 300 times!

31 January GENES

1 A GENE is a section of DNA. Each gene is made up of a UNIQUE SEQUENCE of hundreds, thousands, or millions of chemicals called BASES.

Alleles are responsible for characteristics like hair curliness, eye colour, and number of freckles.

2 Every human has about 20,000 DIFFERENT GENES.
Each gene contains the instructions for a cell to make one of the proteins your body needs to work.

3 Genes determine everything about HOW YOUR BODY IS MADE.
Most genes are the same for everybody because they give the code for things essential for life.

4 Scientists are learning how to REPLACE FAULTY GENES with HEALTHY CODE.
It could help cure genetic diseases like cystic fibrosis.

5 SOME GENES HAVE VARIATIONS, called ALLELES, which affect things such as athletic ability or cause some health conditions.

2 THE BRAIN contains about 86 BILLION NERVE CELLS – more than ten times the number of people on Earth.

It is connected to the rest of the body with long, thin nerve cells called neurons (see p10).

1 THE BRAIN is an organ inside your head that CONTROLS EVERYTHING IN YOUR BODY – from your thoughts and actions, to things like your heartbeat and breathing.

3 THE BRAIN KEEPS WORKING 24 HOURS A DAY.

Each area, or lobe, has a specific task.

PLANNING

THINKING

JUDGING

CREATIVITY

REGULATING EMOTIONS

SPEAKING

MOVEMENT

TOUCH

TASTE

HEARING

SMELL

EMOTIONS

OBJECT RECOGNITION

UNDERSTANDING WORDS

RECOGNIZING FACES

VISION

MEMORY

COORDINATION

The main part of the brain is called the cerebrum.

The cerebellum helps control your body movements.

The brain stem controls many automatic bodily functions.

1

February

YOUR BRAIN

4 The main part of THE BRAIN IS DIVIDED INTO TWO HALVES called the left and right hemispheres.

Each half controls the opposite side of the body.

The spinal cord is a bundle of nerves that connects the brain to the rest of the body.

5 THE OUTSIDE OF YOUR BRAIN HAS MANY DEEP FOLDS AND WRINKLES.

They increase the surface area, allowing more information to be processed.

2 February COPPER

1 COPPER IS A METAL WITH A DISTINCTIVE REDDISH COLOUR.

It turns green when it is exposed to air and water.

2 COPPER QUICKLY TAKES IN HEAT AND HOLDS ON TO IT.

This makes it useful in heating systems, pans, and some computer parts.

3 COPPER IS USED TO MAKE ELECTRICAL WIRES.

This is because it is soft, easy to shape, and conducts electricity easily.

4 THE STATUE OF LIBERTY IS COATED IN A LAYER OF COPPER weighing around 28,000 kg (62,000 lb).

That's around the same as two double-decker buses.

5 COPPER CAN BE COMBINED with other metals TO MAKE ALLOYS.

Mixing it with tin to make bronze or zinc to form brass makes it harder and stronger.

The flame in the torch is coated with 24-carat gold.

The copper has turned green due to a process called oxidation.

3 February MERCURY

1 Mercury is a ROCKY PLANET.

This means it is made mainly of rock and metal, and it has a solid surface.

2 Being the CLOSEST PLANET TO THE SUN makes Mercury very hot.

In the daytime it can reach a sweltering 430°C (800°F).

3 Mercury has very LITTLE ATMOSPHERE.

This is because any gases around it are blasted into space by the Sun.

Mercury's surface is pockmarked with tens of thousands of impact craters from asteroids.

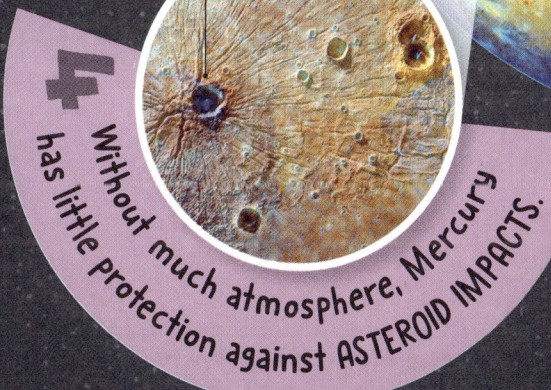

4 Without much atmosphere, Mercury has little protection against ASTEROID IMPACTS.

5 A solar day (sunrise to sunrise) on Mercury is equal to 176 EARTH DAYS.

This is twice as long as its year (88 Earth days) because Mercury spins so slowly on its axis!

4
February
ANTENNAE

1
ANTENNAE COME IN A HUGE RANGE OF SHAPES AND SIZES.
Some are long and thin, others are short and comblike.

2
Lobsters and crabs have TWO SETS OF ANTENNAE.
They have one pair of long antennae and another pair of shorter, branching antennae called antennules.

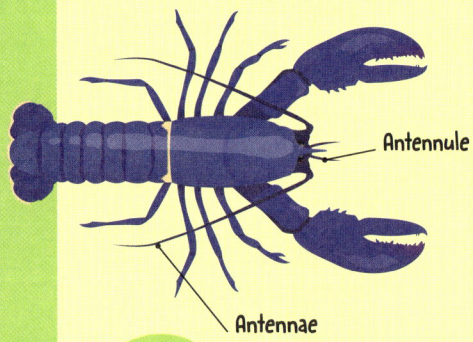

Antennule

Antennae

3
HONEYBEES use their antennae to DETECT CARBON DIOXIDE.
If levels of the gas inside the hive get too high, they beat their wings fast to move more air into and around the hive.

4
ANTS use their antennae TO COMMUNICATE with each other.
They touch antenna to antenna to pass along chemical messages about food sources and threats to their nest.

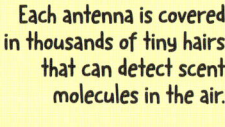

Many male moths have large comblike antennae that they can spread out wide.

Each antenna is covered in thousands of tiny hairs that can detect scent molecules in the air.

5
ANTENNAE ARE PAIRS OF LONG, THIN SENSE ORGANS that insects and some other animals have on their heads.
They can feel the world around them and pick up vibrations, tastes, and smells.

1 Humans collect visual information through their eyes. Eyes are BALLS OF CLEAR, JELLY-LIKE FLUID THAT ALLOW LIGHT TO PASS THROUGH THEM.

2 A ring of coloured muscle called THE IRIS CONTROLS HOW MUCH LIGHT ENTERS THE EYE through the pupil.

In bright light, the pupil shrinks to just 2 mm (0.08 in) wide.

The visual cortex – an area at the back of the brain (see p30) processes images received from the eyes.

Information travels through the optic nerve to the brain, which flips it the right way round.

Lens

Iris

Pupil

Retina

Optic nerve

Light bounces off objects and enters the eye.

The cornea focuses light before it gets to the lens, and helps to protect the eye.

The retina processes the image, but it is upside down at this point.

3 A LENS AT THE FRONT OF THE EYE HELPS TO FOCUS LIGHT.

It becomes rounder when focusing on objects nearby and flatter when viewing those further away.

5 February
VISION

4 At the BACK OF THE EYE is a LIGHT-SENSITIVE AREA of tissue called THE RETINA.

It turns light into electrical signals that are then transmitted to the brain.

5 CONE CELLS in the retina are what DETECT COLOUR.

Humans have three different types of cone cells, but some animals have just one and only see limited shades.

6 February
FOSSILS

1
FOSSILS are the preserved REMAINS of DEAD ANIMALS AND PLANTS that were buried in sand and mud thousands of years ago.

2
PALAEONTOLOGISTS are scientists who STUDY FOSSILS to learn about plants and animals of the past, SUCH AS DINOSAURS.

3
Fossils are made up of the HARD PARTS OF AN ORGANISM.

Soft tissues rot away quickly. Hard body parts, like bones, teeth, and shells, turn to rock as minerals seep into them.

4
POO CAN BE FOSSILIZED! Fossilized poos are called COPROLITES and scientists found one that was over 30 cm (1 ft) long from a TYRANNOSAURUS REX.

5
TINY CREATURES have been found perfectly PRESERVED IN AMBER, a hardened tree resin that OOZES from bark.

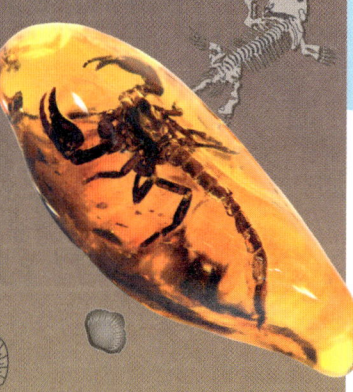

7 February
SNOW

1
When tiny ICE CRYSTALS IN CLOUDS stick together around specks of DUST OR POLLEN in the air, they form SNOWFLAKES.

The shape of each snowflake depends on the temperature and moisture of the air around it as it forms.

2
Snowflakes look white because LIGHT REFLECTS OFF THEM.

They're not really white but translucent (they allow light through).

3
It takes about ONE HOUR for a snowflake to fall to Earth from its cloud.

4
Snow that falls during a thunderstorm is called "THUNDERSNOW".

It only happens in extremely cold conditions.

5
Every year, ONE SEPTILLION snowflakes fall on Earth.

That's 1,000,000,000,000,000,000,000,000 snowflakes!

1 Bridges must be STRONG AND FLEXIBLE to support huge loads and survive strong winds and temperature changes.

Most modern bridges are made from strong, supple steel and concrete.

The Golden Gate Bridge in San Francisco, USA, is a suspension bridge.

2 There are SIX DIFFERENT BASIC TYPES OF BRIDGE.

These are: beam, truss, arch, suspension, cantilever, and cable-stayed.

3 Bridges have to BALANCE IMMENSE FORCES.

These forces include gravity (see page 75), tension (stretching), and compression (squashing).

4

In suspension bridges, the road is HUNG FROM STRONG CABLES, SUPPORTED BY TWO OR MORE TOWERS.

The compression from the weight of the road (and the vehicles driving on it) is transferred through the tight cables, through the towers, and then into the ground.

Suspenders under tension

Anchorage points under tension

Cables under tension

Cables under tension

Cables under tension

Anchorage points under tension

Towers under compression

5 All bridges are designed to SWAY SLIGHTLY IN THE WIND TO STOP THEM SNAPPING.

But when London's Millennium Bridge opened in 2000, it swayed so much that people felt sick, so it had to be closed and adjusted.

8

February
BRIDGES

1 ANIMALS are living things that MOVE AROUND and EAT OTHER LIVING THINGS.

This is different to plants, which make their own food (see p136) and can't move around.

2 Animals can be divided into TWO MAIN GROUPS: VERTEBRATES (animals with a backbone) and INVERTEBRATES (animals without a backbone).

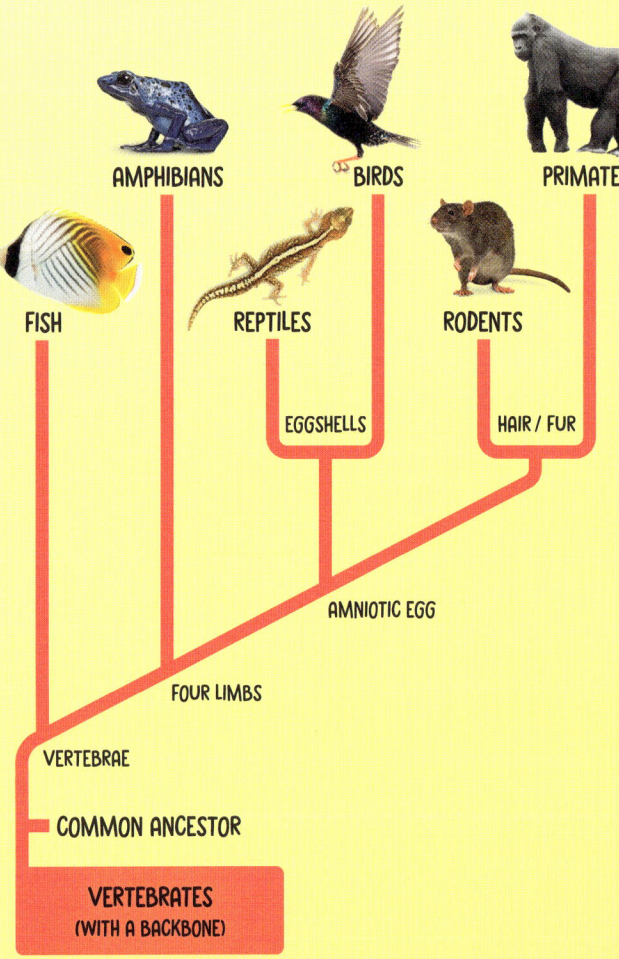

AMPHIBIANS

BIRDS

PRIMATES

FISH

REPTILES

RODENTS

EGGSHELLS

HAIR / FUR

AMNIOTIC EGG

FOUR LIMBS

VERTEBRAE

COMMON ANCESTOR

VERTEBRATES
(WITH A BACKBONE)

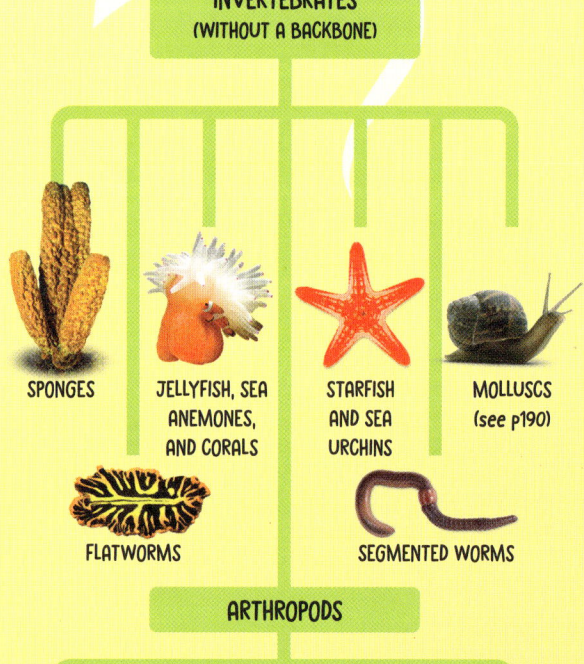

INVERTEBRATES
(WITHOUT A BACKBONE)

SPONGES

JELLYFISH, SEA ANEMONES, AND CORALS

STARFISH AND SEA URCHINS

MOLLUSCS (see p190)

FLATWORMS

SEGMENTED WORMS

ARTHROPODS

SHELLFISH AND WOODLICE

SPIDERS, SCORPIONS, TICKS, AND MITES

INSECTS

MILLIPEDES AND CENTIPEDES

3 Some animals species are ENDANGERED, meaning there are ONLY A FEW LEFT ON EARTH, AND THEY COULD SOON BE EXTINCT (see p160).

For example, the Bengal tiger is endangered due to illegal hunting.

4 There are more than 1.3 MILLION species (unique types) of ANIMAL on EARTH today.

This is less than 1 per cent of the number of species that have ever lived on Earth.

5 Scientists think there could be MILLIONS MORE ANIMAL SPECIES YET TO BE DISCOVERED.

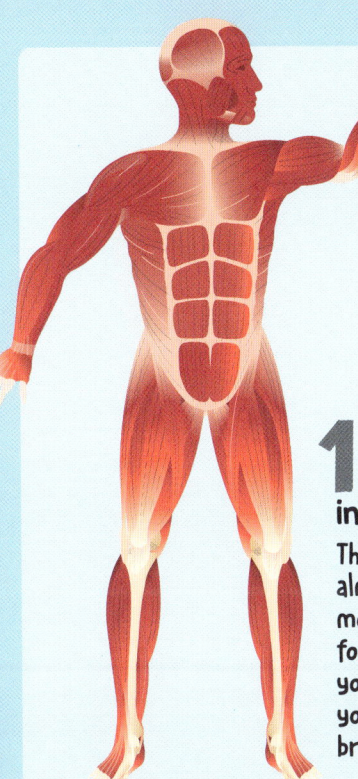

There are 24 different muscles in each arm – all of which are involved when you lift your hand to wave.

1 There are more than 600 MUSCLES in the human body.

They are responsible for almost all of the body's movements – from focusing the lenses in your eyes to lifting your ribs when you breathe in (see p21).

2 There are THREE TYPES OF MUSCLE: cardiac, skeletal, and smooth.

Cardiac muscle makes up the heart, skeletal muscle produces movement, and smooth muscle is found in blood vessels and organs. Skeletal muscle is the only type you can control.

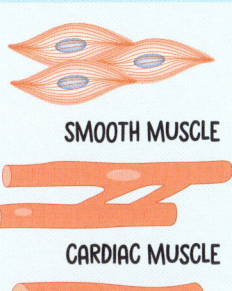

SMOOTH MUSCLE

CARDIAC MUSCLE

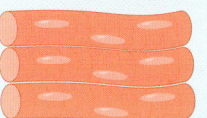

SKELETAL MUSCLE

3 CARDIAC MUSCLE is the only muscle in the body that NEVER GETS TIRED.

It will contract repeatedly every day without pausing for your entire life.

The muscles surrounding each chamber of the heart contract rhythmically in turn to pump blood around the heart and out to the body and lungs.

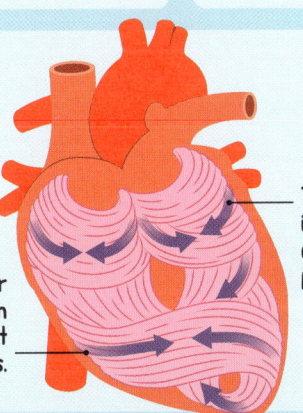

The heart is a powerful muscular pump.

5 Skeletal muscle is made from BUNDLES OF TINY CYLINDRICAL FIBRES.

There are hundreds to thousands of fibres in a muscle, and each fibre can be as little as 0.02–0.08 mm (0.0008–0.003 in) wide.

Your triceps muscle contracts to pull your forearm straight.

The biceps muscle contracts to pull your forearm up.

The triceps muscle relaxes.

4 MUSCLES CAN ONLY PULL (contract), NOT PUSH!

For this reason, they often work in pairs around joints (see p162). For example, to bend your arm at the elbow, your biceps contracts while your triceps relaxes.

Each skeletal muscle has long bundles of fibres inside it that all work together.

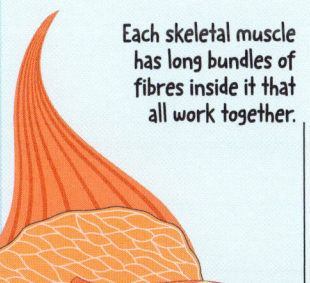

11 February
GREENHOUSE EFFECT

1 Life on Earth needs SOME greenhouse gases in the atmosphere TO KEEP THE PLANET WARM.

2 GREENHOUSE GASES ABSORB and TRAP the SUN'S RADIATION, in the same way a greenhouse keeps the plants inside it warm.

3 Greenhouse gases include CARBON DIOXIDE, METHANE, WATER VAPOUR, OZONE, and NITROUS OXIDE.

One third of the Sun's energy is reflected back into space.

Some heat given off by the Earth's surface escapes into space.

Greenhouse gases trap the rest of the heat in Earth's atmosphere.

4 HUMAN ACTIVITY such as BURNING FOSSIL FUELS and INTENSIVE FARMING creates extra greenhouse gases, which is leading to GLOBAL WARMING.

Sunlight that hits Earth's surface is a mixture of radiation – visible light, ultraviolet, and infrared.

Sunlight that reaches Earth's surface is converted to heat.

5 The greenhouse effect we are currently experiencing is the FIRST TIME Earth's CLIMATE has CHANGED due to HUMAN ACTIVITY.

1
Some animals HIDE IN PLAIN SIGHT by MIMICKING OTHER ANIMALS, to scare predators or to catch prey.

2
The spicebush swallowtail butterfly CATERPILLAR CAN MIMIC A SNAKE to startle predators.

It tucks its head underneath, puffs up its body, and reveals false eyes.

The eyespots are also poisonous.

3
The four-eyed frog has a pair of LARGE FALSE EYES... ON ITS BOTTOM!

These make the frog appear bigger than it is.

12 February MIMICRY

4
A wild cat called THE TREE OCELOT USES VOICE MIMICRY to catch its prey.

It makes the sound of a distressed baby monkey to lure adults monkeys within striking distance!

5
The MIMIC OCTOPUS is the MASTER OF DISGUISE. It can imitate other sea creatures, such as crabs, jellyfish, sea snakes, and shells.

1
BACTERIA ARE TINY LIVING THINGS.

They are so tiny that we can only see them with a microscope, as their bodies are made up of just one cell.

13 February BACTERIA

2
Bacteria live in almost EVERY ENVIRONMENT ON EARTH – even in Arctic snow and deep in Earth's crust.

Harmless bacteria help us break down food and release nutrients. A few, called pathogens, can make us unwell.

3
Bacteria are one of the OLDEST LIFE FORMS on Earth.

They appeared more than three billion years ago – long before the dinosaurs.

4
Bacteria are used TO MAKE SOME FOODS, including cheese, yoghurt, and kimchi.

Flagellum – tail for swimming

Cell wall

DNA – code for life

Pili – for attaching to other surfaces

5
OUR BODIES CONTAIN huge numbers of USEFUL BACTERIA.

14 February
LIFE THAT GLOWS

1 Living things that use chemical reactions to CREATE AND EMIT LIGHT are described as BIOLUMINESCENT.

2 FIREFLIES produce FLASHES OF LIGHT at the end of their abdomens TO ATTRACT A MATE.

3 When DISTURBED BY WAVES, MICROSCOPIC ORGANISMS called dinoflagellates can cause THE OCEAN TO GLOW.

4 76 PER CENT of all ANIMALS in the OCEANS are bioluminescent.

5 Some animals, such as COMB JELLYFISH, emit light when touched, TO STARTLE PREDATORS. Whereas ANGLERFISH produce light TO LURE PREY.

15 February
X-RAYS

1 X-RAYS are a form of RADIATION. They are an invisible wavelength of light on the electromagnetic spectrum (see p204), which includes visible light and microwaves.

2 X-rays travel easily through SOFT TISSUE but not HARD BONE, so they are useful in MEDICINE for seeing inside the body.

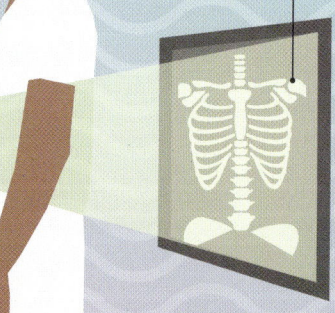

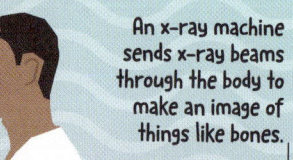

An x-ray machine sends x-ray beams through the body to make an image of things like bones.

3 When it was first discovered, x-ray imaging was used in SHOE SHOPS to check the FIT OF SHOES!

4 AIRPORTS take x-ray images of your BAGS to check for DANGEROUS or FORBIDDEN items.

5 Very hot things give off x-rays. Astronomers use x-ray telescopes to study things in space, such as distant stars and this supernova remnant (see p24).

1 Black holes are huge concentrations of matter packed into a very tiny space. THEY ARE ONE OF THE MOST MYSTERIOUS OBJECTS IN SPACE!

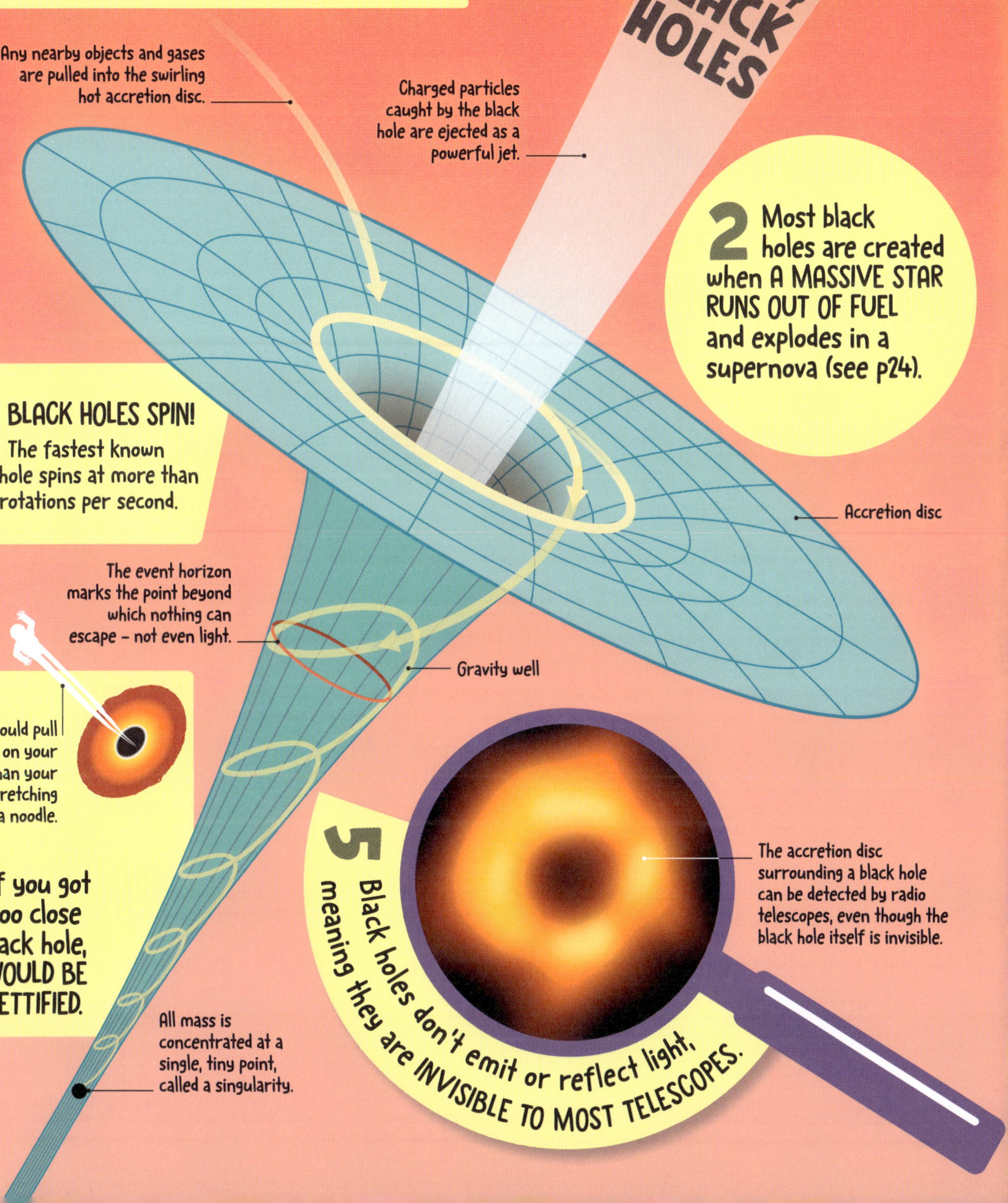

Any nearby objects and gases are pulled into the swirling hot accretion disc.

Charged particles caught by the black hole are ejected as a powerful jet.

2 Most black holes are created when A MASSIVE STAR RUNS OUT OF FUEL and explodes in a supernova (see p24).

Accretion disc

3 BLACK HOLES SPIN! The fastest known black hole spins at more than 1,000 rotations per second.

The event horizon marks the point beyond which nothing can escape – not even light.

Gravity well

Gravity would pull harder on your feet than your head, stretching you like a noodle.

4 If you got too close to a black hole, YOU WOULD BE SPAGHETTIFIED.

All mass is concentrated at a single, tiny point, called a singularity.

5 Black holes don't emit or reflect light, meaning they are INVISIBLE TO MOST TELESCOPES.

The accretion disc surrounding a black hole can be detected by radio telescopes, even though the black hole itself is invisible.

17
February
COMETS

1 Comets are GIANT, DUSTY SNOWBALLS that orbit the Sun.

Near to the Sun, a comet heats up, releasing gases and dust that form a tail millions of kilometres long.

2 If you're lucky, you could see Halley's Comet TWICE IN YOUR LIFETIME.

It can be spied in the sky with the naked eye every 75–76 years. It was last seen in 1986 and will return to our skies again in 2061.

3
COMETS ACTUALLY HAVE TWO TAILS.
Looking closely, astronomers have found that one tail is bright blue and glowing, and the other is white.

The blue tail of the comet is straight and made of electrically charged gas (ions).

The white tail is made of dust and forms a curved path behind the comet.

4
The frozen nucleus or "head" of a comet is ABOUT THE SIZE OF A SMALL TOWN.

It is made of ice, dust, and rock left over from the birth of the Solar System.

5
Comets can be named after the PEOPLE OR SPACECRAFT WHO SPOT THEM.

Comet Hyakutake was discovered in 1996 by Japanese amateur astronomer Yuji Hyakutake using a pair of binoculars.

18
February
MIRRORS

MIRRORS
MIRRORS

The light rays bounce off the mirror at exactly the same angle as they hit it.

MIRROR

3

When you see an object in a mirror, the object APPEARS to be on the OTHER SIDE OF THE MIRROR.

We call this image a virtual image, because it is not real, but an illusion.

1 A MIRROR is a very SMOOTH, REFLECTIVE surface.

Mirrors are usually made of glass with a thin layer of silver on the back to reflect the light. You cannot see an image in rough surfaces because the light that hits them is scattered in many different directions.

2 RAYS of light normally travel in STRAIGHT LINES.

Mirrors work by bouncing light rays from an object into your eyes. This is called reflection.

OBJECT

Light reflected off the flowers travels towards the mirror.

VIRTUAL IMAGE

4

STILL WATER works LIKE a NATURAL MIRROR. Its SMOOTH SURFACE is REFLECTIVE, and can show a reflected image of the sky or a nearby landscape or cityscape.

5

DIAMONDS SPARKLE because their surfaces are ANGLED so that any light that shines into the stone is REFLECTED OFF THE SURFACES inside and out through the top.

Insects like bees pollinate (see p131) flowers when they visit them to collect nectar.

1 Flowering plants produce fruits to PROTECT and SPREAD their SEEDS.

After pollination, an apple flower's ovary begins to swell.

2 Fruits are the MATURE, RIPENED OVARIES of a FLOWER.

They form when a flower has been fertilized (see p131).

3 Not all things we think of as fruits are "TRUE" FRUITS.

Apples and strawberries get their flesh not just from the ovary of the flower, but other parts too. They are known as false fruits or accessory fruits.

The flower and sepals will wither and drop away.

Ovary wall

19

February FRUIT

4 Many fruits are BRIGHT, SWEET, and EDIBLE to attract animals.

The animals eat the fruit and spread the seeds in their droppings.

The flesh of the fruit gets sweeter as it ripens.

The ovules inside the ovaries become seeds for new plants.

The flesh develops from the base of the flower (the receptacle) as well as the ovary.

5 Some fruits are HARD, DRY, or PAPERY.

Sycamore trees have fruits with papery wings that help them to glide away on the wind (see p192).

44

20 February
LAKES

1
A lake is a BODY OF WATER surrounded by LAND.

They usually have areas so deep that sunlight can't reach the bottom.

2
There are more than 304 MILLION LAKES ON EARTH. They form when WATER COLLECTS in a low-lying area.

3
The GREAT LAKES on the US-Canadian border contain 21 PER CENT of the world's FRESH WATER.

4
SUBGLACIAL LAKES beneath the ice in Antarctica may have been ISOLATED from the outside world for 35 MILLION YEARS.

5
Despite its name, the CASPIAN SEA is the LARGEST LAKE on Earth.

Its surface is about the same size as Japan!

21 February
SOLAR ENERGY

1
Waves of ENERGY constantly flow FROM THE SUN towards Earth, WARMING the planet.

2
Stars like the Sun GENERATE ENERGY in their core by NUCLEAR FUSION (see p181).

3
We can CAPTURE this ENERGY using solar panels.

They convert the Sun's heat and light energy to electrical energy.

The solar cell is a dark colour, helping it to absorb sunlight.

Sunlight falls on the cells of the solar panel.

The solar energy makes electrons flow, creating an electric current (see p25).

4
Solar panels are not yet that EFFICIENT – they can only CONVERT around 15–20 PER CENT of the energy they receive.

5
The WORLD'S BIGGEST SOLAR FARM, in Xinjiang province, China, could POWER A LARGE CITY the size of London.

1 Clouds are collections of TINY WATER DROPLETS.

When air cools, the water vapour it carries turns into liquid water, forming a cloud.

CIRRUS

CUMULONIMBUS

The top of a cumulonimbus cloud flattens out, giving it an anvil shape.

2 There are many DIFFERENT TYPES of cloud.

They are divided into groups based on their shape and how high in the sky they are.

CIRROCUMULUS

CIRROSTRATUS

3

THUNDERCLOUDS are called cumulonimbus. They can produce magnificent STORMS with HAIL, LIGHTNING, and THUNDER.

They are tall and towering, reaching up to 20,000 m (65,000 ft) into the sky.

ALTOSTRATUS

Dark grey nimbostratus clouds can produce continuous rain.

NIMBOSTRATUS

ALTOCUMULUS

4

There are clouds on other PLANETS, too – but they are NOT all formed from WATER VAPOUR.

STRATOCUMULUS

Jupiter has clouds made of ammonia mixed with water.

CUMULUS

STRATUS

Fog can be very thick, and forms over the sea as well as over land.

5 FOG is a type of cloud that forms at GROUND LEVEL, not in the sky.

FOG

23
February
DESERT LIFE

1 Deserts are the DRIEST PLACES ON EARTH (see p17). Animals or plants that live there have evolved ways of surviving.

2 Some SCORPIONS AND RATTLESNAKES HIDE underground BY DAY and COME OUT to HUNT AT NIGHT to AVOID the HEAT.

3 A CAMEL'S HUMP contains FAT THAT IT CAN LIVE OFF when there is little food.

A fennec fox's pale fur reflects sunlight and its large ears lose excess heat.

4 Desert plants grow FAR APART to AVOID SHARING WATER.

They often have small leaves or waxy skin to reduce water loss, and long roots to reach water deep underground.

A cactus can swell to collect and store water when it rains.

5 Fog-basking beetles COLLECT WATER from the MORNING FOG on their BODIES.

The beetle does a headstand and the water droplets roll down to its mouth.

24
February
SKULL

1 THE SKULL isn't just one bone – it is actually MADE UP OF 22 BONES.

2 The skull contains HOLLOW SPACES NEAR THE NOSE called sinuses.

They make mucus that traps bacteria and dirt.

3 BABIES have GAPS BETWEEN the SKULL BONES called FONTANELLES.

This allows space for their brains to grow rapidly.

4 OUR SKULL IS MUCH LARGER than that of our APE RELATIVES.

Human brains have tripled in size over the last 3 million years.

5 THE LOWER JAW is the only part of the skull that can MOVE.

Fixed joints called sutures connect the bones of the skull.

Eye socket

The mandible is attached to the skull at moveable joints, similar to a swing.

The mandible (lower jaw) is the largest and strongest bone in the face.

1 Objects get up into the air and are ABLE TO FLY BY GENERATING LIFT.

Birds flap their wings to do this, whereas aircraft have static wings or use spinning rotors.

2 The WINGS OF A PLANE have a shape called AN AEROFOIL.

Air flows faster over the top of the wing.

Wings are tilted at a specific angle to direct the air around them in a way that produces lift.

Air flows slower under the wing, creating higher air pressure that pushes the plane upwards.

3

FOUR FORCES NEED TO BE BALANCED for objects to fly at a constant speed.

The upward force of lift balances out the plane's weight, and thrust from the engine balances out air resistance.

LIFT is generated by the wings.

The plane's WEIGHT pulls it towards the ground.

THRUST is generated by the plane's engines.

As the plane moves through the air, the air pushes back against it. This is air resistance, or DRAG.

Aircraft carriers have no space for a long runaway, so have aircraft catapults to help launch planes!

4

PEOPLE WERE TRYING TO FLY CENTURIES AGO.

In the 9th century, Islamic inventor Abbas ibn Firnas made wings out of wood, silk, and feathers, and flew for a short distance!

25 February FLIGHT

5 Planes need A RUNWAY TO ACCELERATE to a HIGH ENOUGH SPEED for LIFT-OFF.

26
February
BIG CATS

1

[T]he FIVE BIG CAT [SP]ECIES – the lion, [ti]ger, jaguar, leopard, [an]d snow leopard – [ar]e part of the [PA]NTHERINAE FAMILY.

2

Amongst the big cats, LIONS ARE THE ONLY SOCIAL ANIMAL.

They almost always live in groups, called prides, of up to 30 adults and their babies.

3

[TI]GER STRIPE [M]ARKINGS ARE [in]dividually UNIQUE, [li]ke a human [fi]ngerprint.

4

[B]LACK PANTHERS [A]RE any big cats [w]ith a BLACK coat.

[A] genetic variation [ca]uses more dark [pi]gment in their fur.

5

[LE]OPARDS ARE [EX]CELLENT CLIMBERS.

[Th]ey drag their prey up a [tr]ee to eat it away from [ot]her hungry critters.

27
February
MICROSCOPES

1

SPECTACLE MAKERS were the first people to design microscopes.

Microscopic view of human blood cells and the bacteria that cause typhoid.

2

MICROSCOPES work by BENDING RAYS OF LIGHT – using many different lenses.

3

REALLY POWERFUL MICROSCOPES magnify images using a STREAM of ELECTRONS (see p53).

4

Cells, bacteria, and DNA were all DISCOVERED USING MICROSCOPES.

A lens near the eyepiece increases the magnification further.

The light rays refract (bend) when they go through each lens.

5

Today's most powerful microscopes can ZOOM IN TO ONE TEN-MILLIONTH OF A MILLIMETRE (the size of an atom)!

Lenses can magnify the object between four and 100 times.

A mirror reflects light towards the specimen on a slide.

28
February
EXOPLANETS

1 EXOPLANETS ARE PLANETS OUTSIDE OUR SOLAR SYSTEM.

They orbit other stars, and come in many forms, such as gas and ice giants, and big and small rocky planets.

GAS GIANTS are like Jupiter and Saturn, but some may be much hotter.

ROCKY PLANETS are the same size as Earth, or smaller, and made of rock and metal.

2 MORE THAN 5,000 EXOPLANETS HAVE BEEN DISCOVERED SO FAR, and billions are thought to be in our galaxy alone!

3 THE FIRST EXOPLANETS were discovered in 1992.

They were two rocky planets named Poltergeist and Phobetor.

SUPER-EARTHS are rocky like Earth, but less dense, and up to 10 times larger.

NEPTUNE-LIKE PLANETS have atmospheres made of hydrogen and helium.

4 IT WOULD TAKE around 6,300 YEARS to travel to THE NEAREST EXOPLANET, Proxima Centauri b. It is 4.2 light years away.

5 Some exoplanets COULD SUPPORT LIFE.

Several planets similar to Earth have been identified that might be habitable to humans.

29
February
LEAP YEARS

1 A leap year is a year that contains one extra day than usual – 366 DAYS INSTEAD OF 365.

2 IT TAKES EARTH 365.242190 DAYS TO ORBIT THE SUN.

We add up those extra hours into a whole extra day every few years.

3 The first leap years were added to the calendar in 46 BCE by ROMAN EMPEROR JULIUS CAESAR.

4 Leap years are usually every four years, but SOMETIMES THEY ARE SKIPPED!

The next leap year to be skipped will be 2100.

5 More than five million people around the world are BORN ON A LEAP DAY (29th February).

1 There are more than 11,000 SPECIES (unique types) of bird in the world.

Each species is adapted to its specific habitat and lifestyle.

A woodpecker can peck 20 times a second at around 24 km/h (15 mph). Its long tongue helps protect its brain from damage.

2 Birds have HOLLOW BONES so that they are LIGHT ENOUGH TO FLY.

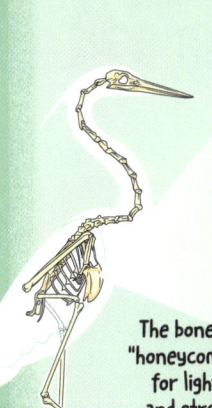

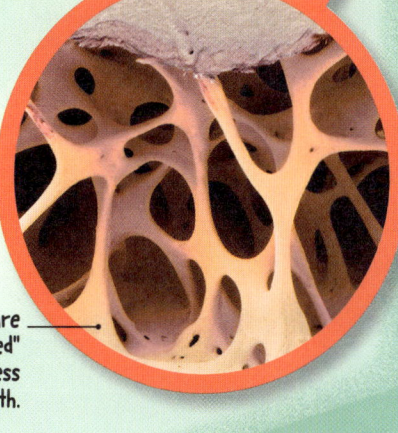

The bones are "honeycombed" for lightness and strength.

3 The DOMESTIC CHICKEN is THE MOST COMMON BIRD species in the world.

The chicken was domesticated from the red junglefowl of Southeast Asia about 8,000 years ago. Male junglefowls have a similar red comb and wattle to cockerels.

1
March
BIRDS

4 FLAMINGOS EAT WITH THEIR HEADS UPSIDE DOWN.

With their heads underwater, they sweep them from side-to-side, filtering algae, brine shrimps, and seeds out of the water.

Their bills are lined with tiny brushes, which allow the water out, while trapping food inside.

5 THE PEREGRINE FALCON IS THE FASTEST BIRD IN THE WORLD.

It can reach speeds of 380 km/h (240 mph) when diving for prey from great heights.

Powerful flight muscles, pointed wings, and stiff feathers combine to allow falcons to fly at great speed.

2
March
URBAN ANIMALS

There are more than 400 million feral pigeons worldwide – most of which live in urban areas.

1
Life in cities ISN'T ALL BAD FOR WILDLIFE.

Cities are noisy and full of traffic and light pollution, but they have lots of food sources, shelter, and fewer predators than rural areas.

Squirrels visit urban parks and gardens for food.

These birds of prey nest on skyscrapers and bridges, which mimic their natural habitat of cliff-faces.

2
The UK capital, LONDON, IS HOME TO more than 15,000 KNOWN SPECIES OF WILDLIFE.

3
NEW YORK CITY, USA, has the world's LARGEST URBAN POPULATION OF PEREGRINE FALCONS.

Coyotes' natural diet is rats, mice, and squirrels, but they will happily eat rubbish, too.

4
On average, EVERY HOME HAS about 100 SPECIES OF CREEPY-CRAWLY!

These range from the spiders and flies we might see, to hidden ones like carpet mites.

Brown rats thrive in city sewers.

5
India's capital, DELHI, IS HOME TO MORE THAN 40,000 MONKEYS.

As their forest habitats have shrunk, rhesus macaques have moved to the city.

1 ATOMS are the microscopic BUILDING BLOCKS OF EVERYTHING on Earth and in space.

2 Atoms contain even smaller subatomic particles called PROTONS, ELECTRONS, AND NEUTRONS.

Each type of atom has a different number of subatomic particles.

Protons are positively charged.

Neutrons have no electrical charge.

3 Gold is an element, which means it is MADE UP OF ALL THE SAME ATOMS.

There are 118 types of atom and therefore 118 elements (see p152).

4 ATOMS ARE FOREVER!

They don't disappear, they just move around the Universe and become part of different substances.

5 THERE'S A LOT OF EMPTY SPACE IN AN ATOM.

If likened to a football stadium, the nucleus would be a pea, and its electrons would be whizzing around at the back of the stands.

Electrons are negatively charged.

Protons and neutrons form the nucleus.

3 March ATOMS

A drop of water contains billions of molecules.

Each water molecule contains three atoms: two hydrogen and one oxygen.

4 March MOLECULES

1 MOLECULES are made of two or more ATOMS that are CHEMICALLY BONDED together.

Although there are only 118 types of atom, together they can form millions of different molecules.

The hydrogen and oxygen atoms share electrons in their outer shells.

O **H**

H

2 WHEN DIFFERENT ATOMS COMBINE, as in a water molecule, they are called "A COMPOUND".

3 The atoms in a molecule chemically BOND together BY SHARING ELECTRONS.

4 The chemical FORMULA for a substance TELLS YOU HOW MANY OF EACH ATOM it has. For example, water is H_2O.

5 Some items we see or use on a daily basis have really long formulas.

For example, the chemical formula for beeswax is: $C_{15}H_{31}COOC_{30}H_{61}$.

5
March
YOUR SLEEP

1 You will spend ONE-THIRD OF YOUR LIFE ASLEEP.

2 While you sleep, your BODY IS STILL WORKING.

Your brain processes and stores information and memories, while your body repairs and replaces cells, and restores energy.

3 Over the course of a night, YOU CYCLE THROUGH FOUR DIFFERENT STAGES OF SLEEP.

Dreams mostly happen during rapid eye movement (REM) sleep, when your brain is most active.

4 Thousands of YEARS OF EVOLUTION has hardwired humans to naturally start to FEEL SLEEPY when it gets DARK, and WIDE-AWAKE in DAYLIGHT.

5 SLEEPWALKING occurs in the DEEP, DREAMLESS SLEEP stage.

The part of the brain that shuts down movement during sleep is woken up, but the rest of you stays deeply asleep.

6
March
SPRINGS

1 A spring is an OBJECT THAT HAS "ELASTICITY".

It can be stretched or compressed and then returned to its original shape.

Compressed spring

A stretched spring has "potential energy" stored inside it until you let it go.

2 Many objects ARE ELASTIC only to a CERTAIN POINT.

If stretched, they deform and can't return to their original shape.

Stretched spring

3 Springs are usually MADE OF METAL BENT INTO A COILED SHAPE.

But a bow that bends to fire an arrow is a spring too!

4 By releasing a stretched spring, its stored energy changes to KINETIC (MOVEMENT) ENERGY.

5 Springs are used in ALL SORTS OF EVERYDAY OBJECTS, from trampolines to mechanical watches.

7
March
VOLCANIC ERUPTIONS

1
Yellowstone Park, USA, sits on top of a SUPERVOLCANO.

Volcanologists keep a close eye on it, as the last time it exploded (640,000 years ago), it sent gas, ash, and molten rock thousands of metres into the air. The volcano then collapsed, sucking in trees, hills, and everything else in an 80-km (50-mile) radius.

2
ERUPTIONS can last for THOUSANDS OF YEARS.

Not all volcanic eruptions are sudden or dramatic. Stromboli, a volcanic island off the Italian coast, has been quietly bubbling away for at least 2,400 years, spitting out lava several times an hour.

3
AT LEAST 20 VOLCANOES ARE ERUPTING WHILE YOU READ THIS PAGE.

There are about 1,500 active volcanoes on Earth, and every year between 70 and 80 of them erupt.

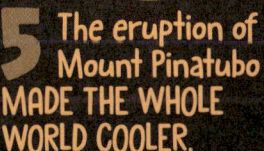

5
The eruption of Mount Pinatubo MADE THE WHOLE WORLD COOLER.

When Mount Pinatubo in the Philippines erupted in 1991, its dust and ash blocked the sunlight and caused a 0.5°C (1°F) drop in temperature worldwide.

4
In 1883, when the volcanic island of KRAKATOA in Indonesia exploded, the BLAST WAS HEARD 4,800 km (3,000 miles) AWAY in MAURITIUS, off the coast of Africa.

8
March
VENUS

1 VENUS IS THE HOTTEST PLANET IN OUR SOLAR SYSTEM - its average surface temperature is a 464°C (867°F).

This is because a dense layer of gases in its atmosphere stops heat from escaping into space.

2 Venus is often described as EARTH'S "EVIL TWIN", because it is similar in size and structure.

But the surface is covered in volcanoes and it is shrouded in a cloak of poisonous gases.

Under the dense, yellow atmosphere, the surface is rocky and volcanic.

The *Venus Express* spacecraft used radar to map the planet's surface.

Venus appears yellow due to its thick blanket of sulphuric acid clouds.

3 A DAY ON VENUS IS 243 EARTH DAYS, because the planet spins really slowly.

This is longer than its year - the planet takes just 225 days to orbit the Sun.

4 VENUS SPINS CLOCKWISE - the opposite direction of Earth and most other planets in the Solar System.

5 Venus is the only planet in the Solar System named after a female god - THE ROMAN GODDESS OF LOVE AND BEAUTY.

9 March SKIN

1 SKIN IS A BARRIER against the world outside your body.

It keeps your insides in, and germs and water out.

2 Your SKIN is the LARGEST ORGAN in your BODY. It makes up about 16 per cent of your total body weight.

3 If you cut yourself, YOUR SKIN REPAIRS ITSELF.

Blood cells clump together to stop the bleeding. A scab then forms over the top while new skin grows underneath.

4 WE ONLY SEE THE TOP PART OF THE SKIN.

But it actually has three layers: the epidermis, the dermis, and the hypodermis.

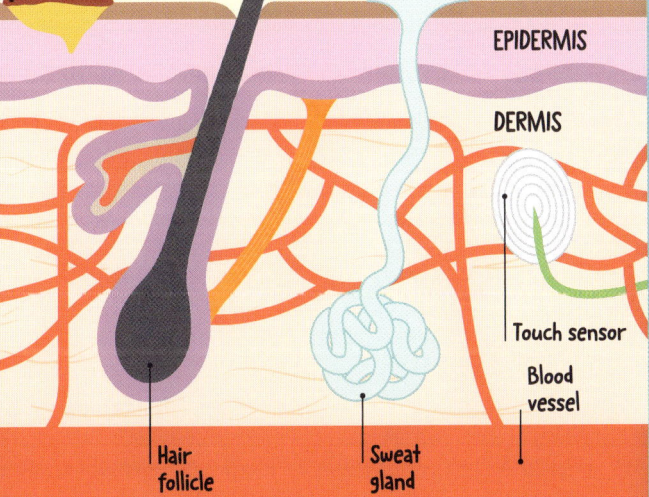

A scab protects the cut while it heals.

Old skin cells flake away.

EPIDERMIS

DERMIS

Touch sensor

Blood vessel

Hair follicle

Sweat gland

HYPODERMIS

5 SKIN IS FULL OF TINY SENSORS, WHICH LET US FEEL TOUCH, HEAT, OR PAIN.

A queen bee can lay 1,500 eggs in just one day.

10 March BEES

1 BEES ARE a group of insects RELATED TO ANTS AND WASPS.

Most bee species live individually, but others, such as honeybees and bumblebees, live in groups called colonies.

2 EACH HONEYBEE HAS A JOB TO DO.

Drones mate with the queen, who lays eggs, while workers collect food and maintain the hive.

3 Bees turn NECTAR from flowers INTO HONEY to feed the colony.

Sugary nectar gives a bee the energy it needs to fly.

4 Bumblebees can DETECT ULTRAVIOLET LIGHT.

This allows them to see seemingly invisible markings on flowers that point them towards nectar.

5 Bees are POLLINATORS.

They carry pollen from flower to flower, which allows the flowers to produce seeds.

EARTHQUAKES

1

EARTH'S CRUST IS MADE UP OF HUGE JIGSAW PIECES called tectonic plates (see p72).

Where two plates meet is known as a fault.

A plate moves one way.

The other plate moves in the opposite direction.

Seismic waves (the earthquake's vibrations) ripple outwards from the focus.

The epicentre is on the surface above the focus.

The focus is where the earthquake begins.

EPICENTRE

FOCUS

PLATE

PLATE

5

THE LARGEST RECORDED EARTHQUAKE HAPPENED IN CHILE IN 1960.

It caused a huge tsunami (see p92), and seismic waves shook the whole world for days afterwards.

A weight is attached to a spring.

Vibrations make the weight move, and the pen records it.

SEISMOGRAPH

2

Earthquakes are the result of SUDDEN MOVEMENT ALONG FAULTS IN THE EARTH'S CRUST – usually where tectonic plates meet.

Larger earthquakes cause cracks in the Earth's crust.

3

About HALF A MILLION EARTHQUAKES happen every year – most TOO SMALL TO FEEL.

4

Networks of sensitive machines called seismographs pick up vibrations from earthquakes and are used to WARN PEOPLE LIVING NEARBY.

12 March CAMOUFLAGE

1 Many animals use CAMOUFLAGE to CATCH PREY or HIDE FROM their PREDATORS.

The sand adder wiggles its tail to look like a tasty worm.

2 Stick insects SWAY GENTLY to pretend to be A BRANCH BLOWING IN THE WIND.

4 Thorn bugs avoid being eaten by LOOKING JUST LIKE THE PRICKLY THORNS OF THE TREES THEY FEED ON.

3 Chameleons, crab spiders, and some frogs can CHANGE COLOUR TO MATCH WHAT THEY'RE SITTING ON.

5 A SAND ADDER knows HOW TO HIDE! It hunts by burying itself in the sand, with just its eyes and tail showing, then pouncing!

13 March HELIUM

1 HELIUM IS THE SECOND LIGHTEST GAS (its atoms weigh very little).

This means it floats upwards, so it is used to fill party balloons. Hydrogen is the lightest gas, but is highly flammable.

Helium is lighter than air, so the balloons float.

Superfluid helium will climb up, out of its container.

2 Helium BOILS and FREEZES at lower temperatures than any other substance KNOWN to SCIENCE.

3 Helium WAS FOUND in SPACE BEFORE it was DISCOVERED on EARTH.

It was first spotted in the Sun's atmosphere in 1868.

4 AT –271°C (–456°F), HELIUM BECOMES A "SUPERFLUID".

At this temperature it becomes a fluid with no friction, and even defies gravity. It can climb walls and squeeze through tiny pores in solid objects.

5 HELIUM MAKES STARS SHINE.

When hydrogen atoms in stars smash together, they produce helium, which releases rays of energy into space.

14
March
SALIVA

1 **SALIVA IS MORE THAN 99 PER CENT WATER.**

The other one per cent is minerals, proteins, and chemicals called enzymes, which begin the process of breaking down food.

2 **DIGESTION STARTS BEFORE YOU EVEN TAKE A BITE!**

Smelling or even thinking about food starts the glands in your mouth producing slimy, enzyme-filled saliva. Your mouth is watering!

3 **DENTISTS LOVE SALIVA.** It kills bacteria in the mouth, which protects your teeth from decay.

4 **SALIVA KEEPS YOUR BREATH FRESH!**

If you sleep with your mouth open, your saliva dries up, making your breath smell in the morning.

5 You produce UP TO 1.5 litres (2.5 pt) OF SALIVA EVERY DAY.

That's 10 litres (17.5 pt) in a week – equivalent to 30 cans of drink – which means you swallow that amount, too!

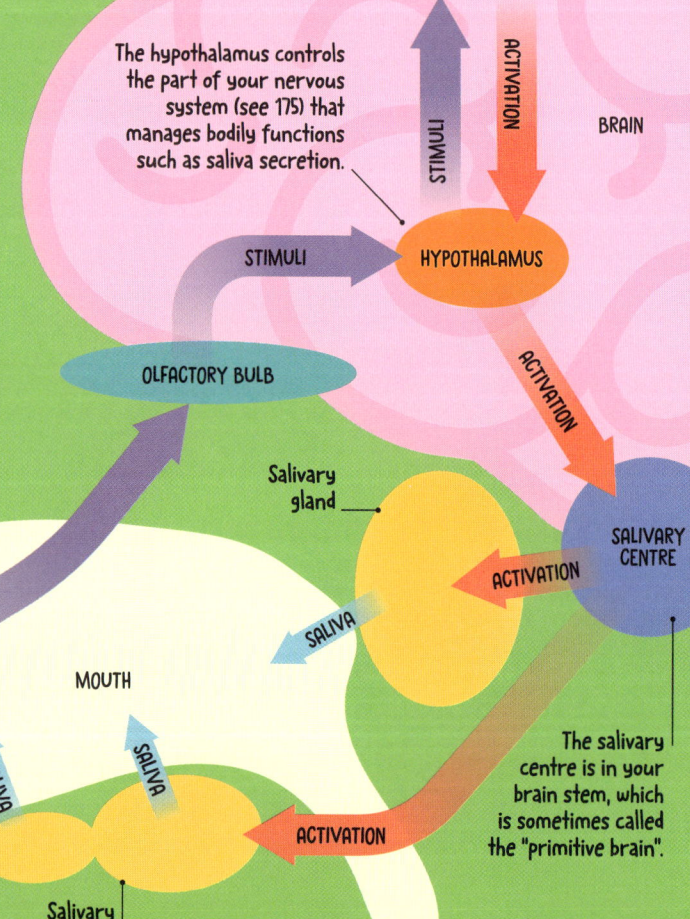

The information about what is causing you to salivate is also sent to higher areas of the brain, including your memory, which recognizes that smell as food.

HIGHER AREAS

The hypothalamus controls the part of your nervous system (see 175) that manages bodily functions such as saliva secretion.

STIMULI

ACTIVATION

BRAIN

STIMULI

HYPOTHALAMUS

ACTIVATION

OLFACTORY BULB

Salivary gland

SALIVARY CENTRE

ACTIVATION

SALIVA

NOSE

MOUTH

STIMULI

The salivary centre is in your brain stem, which is sometimes called the "primitive brain".

SALIVA

SALIVA

ACTIVATION

Salivary glands

15
March
CARBON

1 CARBON IS ESSENTIAL IN OUR WORLD.

It's everywhere, from carbon dioxide in the air to the ink in these words.

2 ALL LIFE ON EARTH IS "CARBON-BASED".

Humans are made of nearly 20 per cent carbon in one form or another.

3 Carbon BONDS EASILY WITH OTHER ATOMS.

It is known to form nearly 10 million compounds (see p53).

4 Carbon is constantly MOVING FROM PLACE TO PLACE in a process known as the CARBON CYCLE (see p86).

The pyramid shape makes diamond hard.

5 DIAMONDS AND GRAPHITE (pencil lead) are both FORMS OF PURE CARBON.

Diamonds are hard and transparent, while graphite is soft and black, because their atoms are arranged differently.

DIAMOND

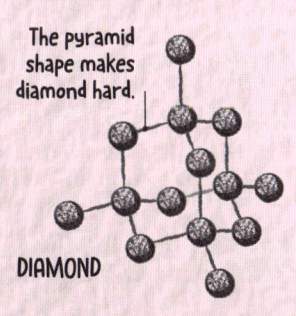

GRAPHITE

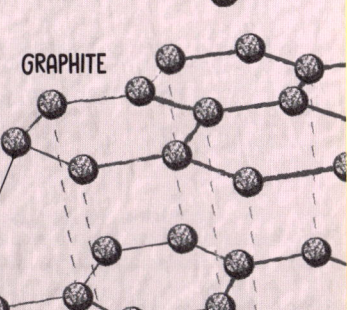

Layers of carbon atoms slip over each other easily.

16
March
ANIMAL SOUNDS

1 VERVET MONKEYS use different WARNING SOUNDS for different PREDATORS.

The troop then know whether to race up a tree or clamber down to safety.

2 THE SPERM WHALE IS THE WORLD'S LOUDEST ANIMAL.

Its clicks are louder than a rocket launch!

3 Frogs of the same species have DIFFERENT ACCENTS IN DIFFERENT LOCATIONS.

4 THE "TWIT TWOO" OWL SOUND IS A DUET.

A male tawny owl sings "twit" and the female responds to him with "twoo".

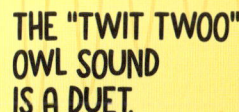

5 ELEPHANTS COMMUNICATE USING INFRASOUND.

This is too low frequency for us to hear, but other elephants can hear it 10 km (6 miles) away.

The balloon has a negative charge.

Opposite charges attract.

The hair has an excess positive charge.

1
If you RUB YOUR HEAD with a balloon, a CHARGE CALLED STATIC ELECTRICITY builds up and makes your HAIR STAND ON END.

2 As you rub, negatively charged ELECTRONS (see p53) TRANSFER FROM YOUR HAIR to the balloon.

This makes the balloon more negatively charged and your hair more positively charged.

3
Objects with OPPOSITE CHARGES ATTRACT each other (see p165), so your hair is pulled towards the balloon.

17
March
STATIC ELECTRICITY

4
LIGHTNING (see p193) occurs when there is a BUILD-UP of static electricity inside a CLOUD.

5
STATIC SHOCKS happen when EXTRA ELECTRONS clinging to your body are SUDDENLY RELEASED by touching something like a metal door handle.

18
March
SALT

1 SALT IS THE MINERAL SODIUM CHLORIDE – a compound of the elements sodium and chlorine.

2 ALL ANIMALS, including humans, NEED SALT TO SURVIVE.

It is used by our muscles and nerves, and helps keep the body's fluids balanced.

3 The salt we use in food is made up of TINY CUBIC CRYSTALS.

It is usually white, but can be different colours if it contains impurities.

4 Salt can be MINED or COLLECTED FROM SEAWATER.

The Sun evaporates the water from shallow pools, leaving salt behind.

5 For thousands of years, salt has been used to PRESERVE AND SEASON FOODS.

But it has many other uses, from helping to remove ice from roads in winter to making fire extinguishers.

19
March
ARTIFICIAL LIMBS

1 ARTIFICIAL LIMBS help people TO MOVE OR HOLD THINGS when they have lost a body part.

They are also known as prosthetic limbs or prostheses.

2 One of the EARLIEST KNOWN PROSTHESES was an ARTIFICIAL TOE.

The 3,000-year-old toe was made of wood.

3 A BIONIC LIMB is an ELECTRONIC artificial bodypart that uses motors and sensors to move the limb.

4 Some bionic limbs use THOUGHTS TO CONTROL MOVEMENT.

They work using sensors attached to nerves in the remaining limb.

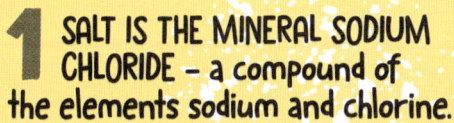

Carbon fibre material makes the prosthetic strong, light, and springy.

5 Running "BLADES" used by athletes were inspired by the legs of KANGAROOS AND CHEETAHS.

The shape of the blade is similar to the hind legs of a cheetah.

20
March
SEAWEEDS

Tall seaweeds have air sacs that help them float upright to reach the sunlight.

1 More than 12,000 SPECIES OF SEAWEED – a marine algae – live in seas, lakes, and rivers.

2 Seaweeds are VITAL in the OCEAN FOOD CHAIN, providing food for many creatures.

3 Seaweeds produce 70 PER CENT MORE OXYGEN THAN LAND PLANTS. They also absorb carbon dioxide.

4 Seaweads range in SIZE from TINY Phytoplankton to TOWERING giant kelp.

5 Seaweeds either float near the surface or grow in shallow waters because they USE PHOTOSYNTHESIS to create food from the Sun, like plants do (see p136).

21
March
AMAZING EGGS

1 Reptiles, insects, spiders, amphibians fish, and birds ALL LAY EGGS.

Apart from birds, all these groups have some species that perform live births.

Skate egg

Reptile eggs are usually soft but tough.

Turtle egg

Frog eggs

Some shark eggs are spiral-shaped!

2 Most animals DON'T lay their eggs IN NESTS.

Some mosquitoes lay eggs in rafts, and some fish release their eggs into the open water.

The female mosquito builds a floating raft made of many eggs stuck together.

3 THE BIGGEST EGG of any animal is the ostrich egg.

The largest recorded was more than 2.5 kg (5.5 lb). That's equal to the weight of 40 chicken eggs!

Frogs' eggs are laid in protective jelly.

4 The American bullfrog CAN LAY as many as 20,000 EGGS IN ONE GO!

5 A kiwi's egg takes up about a QUARTER OF ITS BODY.

22
March
FRICTION

The head of the match and the sandpaper are both rough, causing friction and heat, which ignites chemicals on the head.

PUSHING FORCE

MOTION

1 FRICTION is a force that SLOWS THINGS down when they move AGAINST EACH OTHER.

Rough surfaces have more friction than smooth ones.

FRICTION

All surfaces have some friction, even if they are extremely smooth.

Adding oil can help things slide more easily over each other.

2 MATCHES use friction to MAKE FIRE.

3 The FRICTION between surfaces CREATES HEAT.

Test this out by rubbing your hands together – they will get warmer.

4 Friction ISN'T ALWAYS HELPFUL.

If machine parts rub against each other, they can wear out faster.

5 Car TYRES use FRICTION.

Friction allows a car to grip the road as it accelerates, brakes, and turns corners.

23
March
CLOWNFISH

Anemone stings paralyse other fish, but not the clownfish.

4 Clownfish can CHANGE SEX.

If the strongest female dies, the strongest male changes sex and starts laying eggs.

1 CLOWNFISH live among the VENOMOUS, STINGING TENTACLES of SEA ANEMONES for protection.

2 Sea anemones and clownfish exhibit what's known as "MUTUALISM".

This means they help each other to survive.

3 Clownfish LURE PREY for anemones to EAT.

They don't go far from safety, though, as they aren't strong swimmers.

5 MALE clownfish are GREAT DADS.

They prepare and clean the nest and guard the eggs until they hatch.

24 March
BONES

1 There are two types of bone in the human body: SPONGY BONE and COMPACT BONE.

2 Spongy bone is A LITTLE LIKE HONEYCOMB. It is packed with tiny holes that are filled with blood vessels and red bone marrow.

Spongy bone is usually found at the enlarged ends of long bones, like the femur (thigh bone).

Compact bone forms the hard outer shell of most bones.

Blood vessels provide the bone with oxygen and nutrients.

3 Dense compact bone makes up 80 PER CENT OF THE HUMAN SKELETON.

It provides protection and strength for the bones.

Yellow bone marrow has cells that can develop into cartilage, fat, or bone cells.

4 Billions of blood cells are MADE INSIDE YOUR BONES EVERY DAY.

They are made in the red bone marrow found inside some of your bones (see p122).

5 OLD BONE CELLS ARE ALWAYS BEING REPLACED.

This is why broken bones can heal – over time, new bone cells form at the site of a break to mend the bone.

25 March
TRAINS

1 EARLY TRAINS RAN ON STEAM.

British engineer Richard Trevithick developed the first in 1804.

2 Steam locomotives BURNED COAL TO PRODUCE STEAM, which PUSHED DOWN PISTONS to TURN THE WHEELS.

3 High-speed Japanese BULLET TRAINS HAVE A LONG NOSE AT THE FRONT.

This makes them more aerodynamic.

4 The LONGEST FREIGHT TRAIN was made up of 682 WAGONS, and had EIGHT LOCOMOTIVES PULLING IT!

It stretched for around 7.4 km (4.6 miles) as it travelled across Western Australia.

5 Today's fastest trains can REACH SPEEDS OF 603 km/h (375 mph).

1 Auroras are named after AURORA, the ancient ROMAN GODDESS of the DAWN.

2 AURORAS OCCUR AT BOTH THE NORTH AND SOUTH POLES.

The Northern Lights are known as the aurora borealis, and the Southern Lights are known as the aurora australis.

3 Auroras are caused by electrically charged particles from the Sun, called THE SOLAR WIND.

Earth's magnetic field protects us from the particles, but a few particles sneak in at the poles and collide with gas atoms in our atmosphere and give off light.

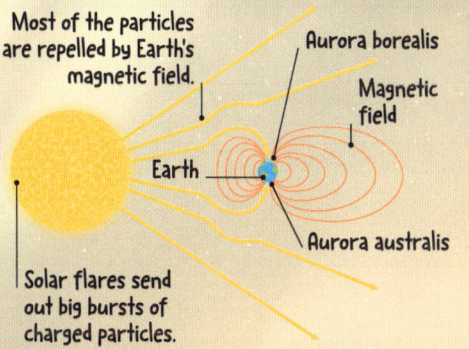

Most of the particles are repelled by Earth's magnetic field.

Aurora borealis

Magnetic field

Earth

Solar flares send out big bursts of charged particles.

Aurora australis

4 Each gas in our atmosphere GIVES OFF A DIFFERENT COLOURED LIGHT when it becomes excited by the Sun's energy.

When the colours mix, they can create pink and yellow lights.

Nitrogen glows blue and purple when excited by the Sun's particles.

Oxygen gives off green and red light when excited by the Sun's particles.

26
March
AURORAS

5 AURORAS also OCCUR ON OTHER PLANETS in our Solar System, including Jupiter, Saturn (right), Neptune, and Mars!

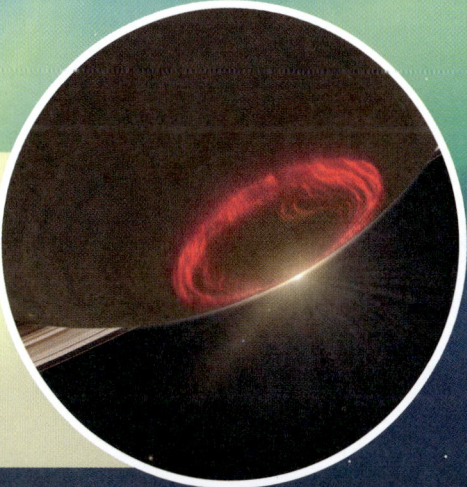

1 Cloning is the process of MAKING IDENTICAL COPIES OF LIVING THINGS.

Some species of plants and bacteria naturally reproduce by cloning.

2

PLANTS CAN BE CLONED BY TAKING A CUTTING and planting it somewhere new.

Roots will sprout from the cutting, beginning its growth into a new plant.

27
March
CLONING

1. A donor cell is taken from the sheep to be cloned.

2. The cell's nucleus is extracted.

3. An egg cell is taken from another sheep and the cell's nucleus is removed.

3

An adult animal can be cloned by FUSING THE NUCLEUS (centre) OF ONE OF ITS CELLS (see p16) INTO the EGG CELL of ANOTHER ANIMAL.

4. The empty egg cell and the donor nucleus are given an electric shock.

5. The two fuse together.

6. The cell multiplies to form an embryo.

7. The embryo is implanted into a surrogate mother.

8. A cloned lamb is born – it is identical to the donor sheep.

4 THE FIRST MAMMAL CLONED was A SHEEP named DOLLY.

Other animals that have been cloned since, including cats, deer, dogs, horses, and rabbits.

5

HUMANS HAVE NOT BEEN CLONED... YET! Cloning humans is banned in most parts of the world.

1

DOGS DESCENDED FROM WOLVES. They were first tamed by humans 20,000-40,000 YEARS AGO.

PUG

Selective breeding over thousands of years created the different breeds we have today.

Their excellent sense of smell makes dogs useful for detecting dangerous and illegal substances.

GREY WOLF

2

Dogs have an INCREDIBLE SENSE OF SMELL.

They have about 250 million scent receptor cells, while humans have just 5 million.

3

Dogs can DETECT HEALTH CONDITIONS and ILLNESSES.

They are able to recognize conditions such as cancer and COVID-19 from changes in the smell of a human's breath alone.

4

Their COLD NOSES help dogs SENSE HEAT!

Unlike most animals, a dog's nose is bumpy, cold, and packed with nerves. This helps them to detect weak heat radiation.

5

Dogs SWEAT THROUGH THEIR PAWS, where they have no fur.

Sweat glands on the paws release sweat to keep cool. They also pant when hot – this allows water to evaporate from their tongue, cooling them down.

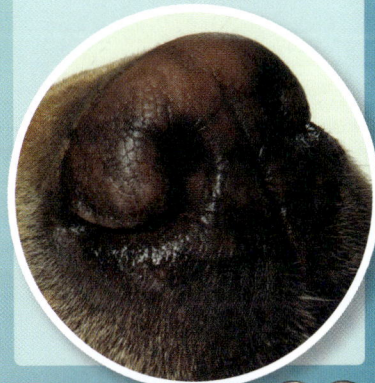

69

1

A helicopter uses ROTATING WINGLIKE ROTOR BLADES to FLY.

As they spin, the rotor blades create a force called lift (see p48) that helps the helicopter move upwards into the air.

2

Helicopters usually have TWO sets of ROTOR BLADES.

The main set on the top lifts the helicopter into the air. A smaller, vertical set attached to the tail stops the 'copter from spinning around.

The rotor blades are powered by an engine. They spin hundreds of times a minute.

The tail rotor is used for steering, and stops the helicopter from spinning out of control.

29
March
HELICOPTERS

The pilot controls the helicopter from the cockpit, using sticks, levers, and pedals that adjust the rotors.

3

Helicopters can HOVER.

To hover, the pilot changes the angle of the rotor blades to produce just enough lift to balance the helicopter's weight, so it hovers in one place.

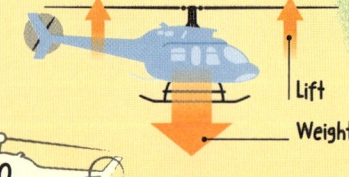

Lift

Weight

4

TILTING the rotors CHANGES a helicopter's DIRECTION.

By changing the tilt of the main rotor blades, a helicopter can move forwards, backwards, and sideways in any direction.

Tilting the rotors forwards moves the helicopter forwards.

5

Helicopters are perfect for SEARCH-AND-RESCUE MISSIONS.

They can take off vertically, which means they don't need a runway, and they can hover over tricky places, such as cities, mountains, and the open ocean.

30
March
CAVES

1

Most CAVES FORM when RAINWATER SEEPS THROUGH CRACKS IN THE GROUND in landscapes made of soft rock, such as limestone, gypsum, and dolomite.

5

HUMANS LIVED IN LAVA CAVES IN THE STONE AGE.

The caves were created when the surface of a stream of lava solidified, but continued to flow beneath, carving out tunnels called lava tubes.

2

The trickling water takes MILLIONS OF YEARS to DISSOLVE the ROCK, until great caverns are formed.

4

The Cave of Crystals in Mexico is full of HUGE, MILKY-WHITE CRYSTALS.

The crystals have been growing for thousands of years and some are over 11 m (37 ft) long and 1 m (3 ft) wide.

3

The LONGEST CAVE SYSTEM in the world is just over 30 TIMES the length of New York's Manhattan Island.

The Mammoth Cave in Kentucky, USA, stretches for over 688 km (426 miles).

1 Earth's crust and upper mantle is BROKEN into LARGE SLABS, known as TECTONIC PLATES.

There are seven major plates, which make up 94 per cent of Earth's surface, and eight minor plates.

The largest slab is the Pacific plate, which is almost 103 million sq km (40 million sq miles).

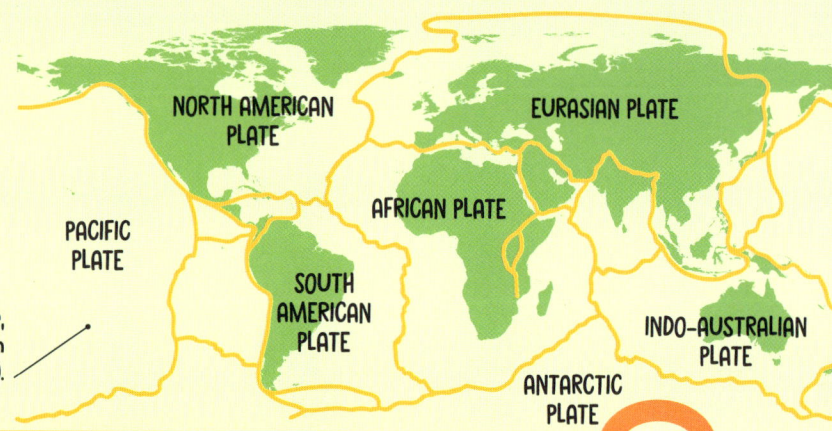

NORTH AMERICAN PLATE

EURASIAN PLATE

AFRICAN PLATE

PACIFIC PLATE

SOUTH AMERICAN PLATE

INDO-AUSTRALIAN PLATE

ANTARCTIC PLATE

2 The tectonic plates are CONSTANTLY MOVING!

They sit on top of the softer lower mantle of the Earth, which churns due to heat from the Earth's core.

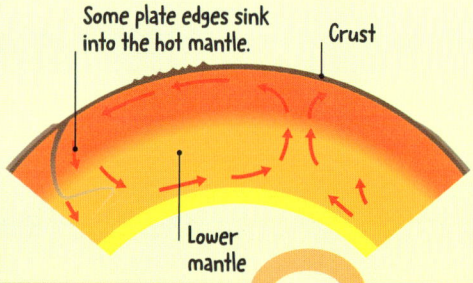

Some plate edges sink into the hot mantle.

Crust

Lower mantle

4 The plates move away, towards, or alongside each other by about 1.5 cm (0.6 in) each year – ABOUT THE RATE YOUR FINGERNAILS GROW!

The two plates slide past each other.

One plate dives under the other.

The plates move apart.

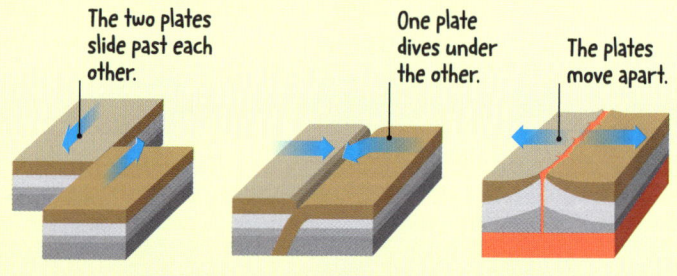

3 The edges of tectonic plates are called "PLATE BOUNDARIES".

The Japanese islands lie where four plates meet, which causes lots of earthquakes.

31 March
TECTONIC PLATES

PACIFIC OCEAN

5 The edge of the Pacific plate is called "THE RING OF FIRE".

About 90 per cent of all volcanoes and 75 per cent of all earthquakes on Earth occur here.

Most of the boundary has one plate diving beneath another.

1 April INTESTINES

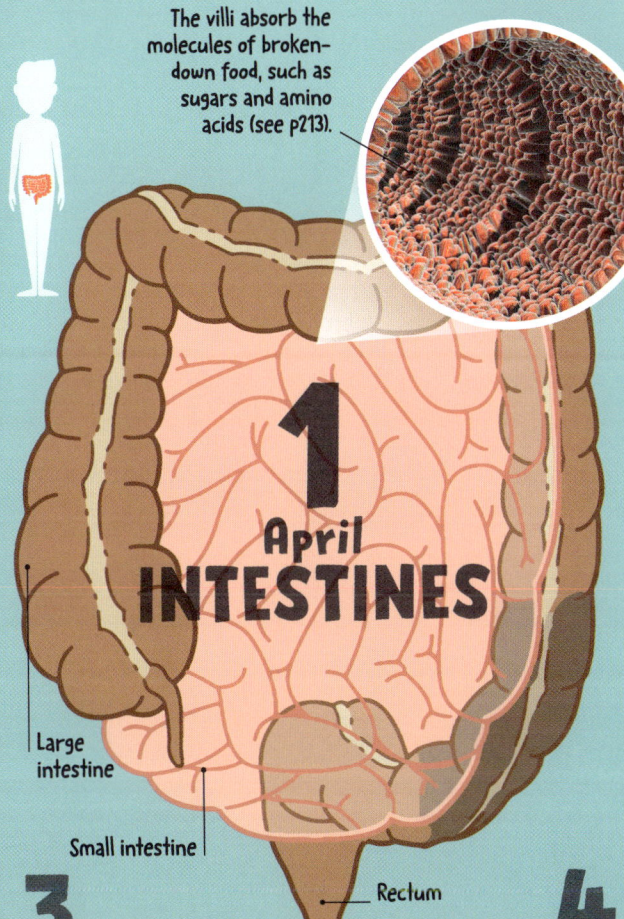

1
THE SMALL INTESTINE IS WHERE MOST OF YOUR FOOD IS DIGESTED.

2
The small intestine IS LINED WITH MORE THAN 5 MILLION VILLI, which look like tiny fingers.

The villi absorb the molecules of broken-down food, such as sugars and amino acids (see p213).

Large intestine

Small intestine

Rectum

3
The small intestine is LARGER THAN THE LARGE INTESTINE!

If you stretched out the small intestine, it would be 6.7 m (22 ft) long!

4
The large intestine CONTAINS UP TO 1,000 DIFFERENT KINDS OF BACTERIA.

Many of these bacteria help you to digest food.

5 The large intestine produces ENOUGH GAS every day TO BLOW UP A PARTY BALLOON.

2 April MOULD

1
MOULD is a type of fungus, and is CLOSELY RELATED TO MUSHROOMS and toadstools.

There are lots of different types of mould, and many of them grow very fast.

2
Mould spreads using TINY SEEDLIKE SPORES, which grow into LONG THREADS CALLED HYPHAE.

The hyphae develop spore capsules, which release new spores, allowing the cycle to start again.

Spores

3
MOULD CAN APPEAR on a damp surface IN JUST 24 HOURS, and can create new seedlike spores after a few days.

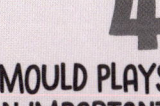

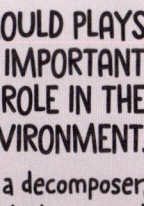

4
MOULD PLAYS AN IMPORTANT ROLE IN THE ENVIRONMENT.

It is a decomposer, which means it feeds on dead things and helps to break them down into soil.

Hyphae

5
The "blue" part of BLUE CHEESE is a TYPE OF MOULD called *Penicillium*.

This mould is safe to eat and does not produce any toxins. Many people think it is delicious!

3
April
YOUR SKELETON

The skull protects the brain.

1

The human skeleton is THE BODY'S FRAMEWORK.

It gives shape and support to the body, protects important organs, and provides a frame for muscles (see p37) to attach to.

2

Bones are held together at JOINTS (see p162) that allow the skeleton to BEND AND TWIST.

The rib cage protects the heart and lungs.

The human spine has 33 bones called vertebrae.

3

More than HALF YOUR SKELETON'S BONES are found in your HANDS AND FEET.

There are 27 in each hand and 26 in each foot.

The femur (thigh bone) is the longest and heaviest bone in the skeleton.

4

BABIES HAVE MORE BONES THAN ADULTS – babies have about 270 bones in their skeletons, while adults have just 206!

This is because some of your bones fuse together as you age.

5

The SMALLEST BONE IN THE SKELETON is just 3 mm (0.12 in) long.

The tiny stapes, or stirrup, bone is found in your ear.

74

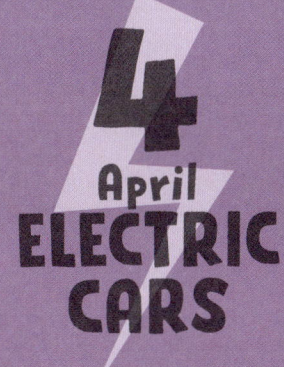

4
April
ELECTRIC CARS

1 ELECTRIC CARS do not use petrol or diesel – instead, they are FUELLED BY A LARGE, RECHARGEABLE BATTERY PACK.

2 The battery pack is made up of THOUSANDS OF SMALL RECHARGEABLE CELLS.

It powers an electric motor, which turns the wheels.

3 Electric cars were FIRST DEVELOPED IN THE 1830s.

They could not travel far before their batteries ran out!

4 Today, it can take just 15 MINUTES TO CHARGE THE BATTERY of some electric car models.

5 The NISSAN LEAF is one of the most POPULAR electric cars.

More than 650,000 have been sold worldwide.

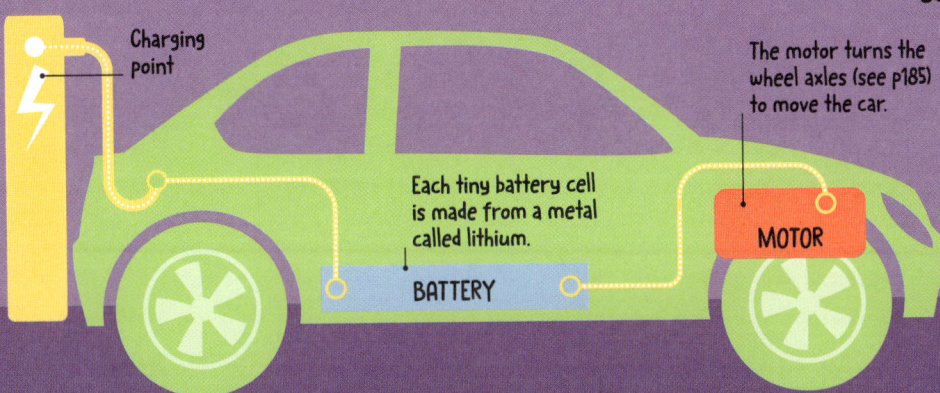

Charging point

The motor turns the wheel axles (see p185) to move the car.

Each tiny battery cell is made from a metal called lithium.

BATTERY

MOTOR

5
April
GRAVITY

1 English scientist Isaac Newton developed the THEORY OF GRAVITY in 1687, after he watched an APPLE DROP FROM A TREE.

Newton wondered why apples fall straight to the ground and realized the same force probably holds us to the ground too.

2 GRAVITY IS THE FORCE by which a planet or other body DRAWS OBJECTS TOWARDS ITS CENTRE.

It holds planets in orbit around the Sun and keeps the Moon in orbit around Earth.

3 The gravitational pull of THE MOON CREATES TIDES ON EARTH.

It pulls on Earth's water, causing it to bulge out from the planet.

4 WEIGHT is based on GRAVITY.

Your weight on Earth is how hard gravity is pulling you towards Earth's surface.

5 BLACK HOLES (see p41) have the GREATEST GRAVITATIONAL PULL in the Universe.

6

April
WIND POWER

1 Wind turbines convert the energy of the wind into ELECTRICAL ENERGY.

Clusters of these turbines are called wind farms.

When the blades of a wind turbine spin, they turn a generator.

2

Most WIND TURBINES have THREE blades.

Fewer blades are more efficient, but turbines with just two blades can wobble too much.

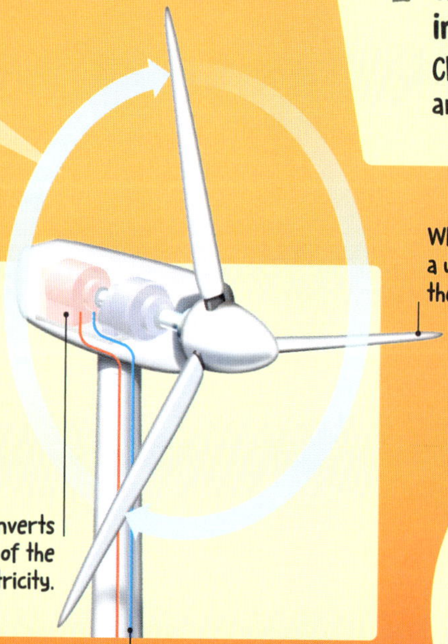

The generator converts the movement of the blades into electricity.

Cables in the tower feed the electricity to a substation, where it can be passed to homes.

3 Many wind farms are BUILT AT SEA rather than on land.

The world's largest offshore wind farm off the coast of the UK has 165 turbines.

5 In high winds, blades on the heaviest wind turbines can SPIN AT SPEEDS of 290 km/h (180 mph)!

4 The TALLEST WIND TURBINES can be more than 185 m (606 ft) – taller than the STATUE OF LIBERTY.

7

April
SMALLEST ANIMALS

1

One of the SMALLEST MAMMALS on Earth is the ETRUSCAN SHREW.

It weighs just 1.5 g (0.05 oz).

2

The SMALLEST BIRD in the world is the BEE HUMMINGBIRD.

Their nests are 2.5 cm (1 in) across and their eggs are smaller than coffee beans!

3

AT JUST 7.9 mm (0.31 in) LONG, *Paedocrypis progenetica* from Sumatra, Indonesia is ONE OF THE SMALLEST known SPECIES OF FISH.

4

The world's SMALLEST AMPHIBIAN is *Paedophryne amauensis* – a fly-sized frog that lives in PAPUA NEW GUINEA. They are just 7.7 mm (0.30 in) long.

Brookesia micra used to be the smallest chameleon until an even tinier one was discovered in 2021.

5

The world's SMALLEST REPTILE is *Brookesia nana* – a chameleon the size of a SUNFLOWER SEED.

8

April
SUBMARINES

With air filling its ballast tanks, the submarine floats on the surface of the water.

1 Submarines are vessels that can operate COMPLETELY UNDERWATER.

Most are used for military purposes.

2 Some models can DIVE TO DEPTHS of 450 m (1,500 ft) below the surface of the water.

3 A submarine can dive quickly by PUMPING WATER into large BALLAST TANKS.

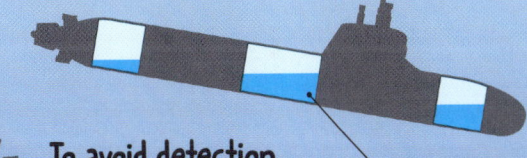

4 To avoid detection, submarines have QUIET ENGINES and can usually be DISCOVERED ONLY BY SONAR (see p119).

To dive, the submarine fills its ballast tanks with water, making it heavier.

To rise back up, the submarine pumps air into the ballast tanks.

5 There are thought to be more than 485 submarines CURRENTLY OPERATING globally.

1 Amphibians evolved from a PRIMITIVE FISH nearly 400 MILLION YEARS AGO.

Their ancestors were the first four-legged animals to leave water and climb onto land.

Amphibians' fish ancestor may have been similar to modern lungfish.

2 FROGS, TOADS, NEWTS, SALAMANDERS, and CAECILIANS are all types of amphibian.

Over 7,700 out of the 8,700 species of amphibian are frogs and toads.

Poison-dart frogs secrete poison from their skin. Their bright colours warn predators they are poisonous.

Caecilians look like large worms.

3 CAECILIANS are amphibians with NO LEGS.

They live underground where it is too dark to see, so many caecilians have no eyes.

Adult frog

Adults breed and lay eggs, called frogspawn.

4

Most amphibians go through "COMPLETE METAMORPHOSIS" (see p227).

This means that the adults change to look totally different from their young.

Frogspawn becomes tadpole

Young frog, known as a froglet

Tadpole grows rear legs

Tail shortens and front legs appear

Axolotl is an Aztec word meaning "water dog".

5 If an AXOLOTL LOSES A LIMB, it can GROW IT BACK.

This Mexican salamander never grows up. It keeps its gills as an adult and stays living in water.

9

April
AMPHIBIANS

10 April
CLIMATE CHANGE

1 Earth's climate is ALWAYS CHANGING, but we are now in a period of RAPID CLIMATE CHANGE caused by HUMAN ACTIVITY.

2 The rapid climate change is being caused by POLLUTION of the ATMOSPHERE.

Human activity, such as farming and burning fossil fuels (see p212) for homes, industry, and transport, increases the greenhouse gases in the atmosphere (see p38).

3 The Arctic POLAR ICE CAP has REDUCED in size by 50 PER CENT since the year 2000.

4 If all the ICE SHEETS AND GLACIERS MELTED, sea levels would rise by over 60 m (195 ft).

All coastal cities on Earth would be submerged.

5 CUTTING TREES to clear land for farming is ADDING TO THE PROBLEM.

Trees remove carbon dioxide from the atmosphere, so clearing trees increases the greenhouse effect.

11 April
STARS

1 Stars are GIANT LUMINOUS BALLS OF HOT GAS.

They are made mostly of the gases hydrogen and helium.

Blue stars are hottest
BLUE
30,000°C (54,000°F)

2 The BRIGHTEST STAR in the night sky is SIRIUS.

It is a pale blue-white star about twice as massive as our Sun.

DEEP BLUE-WHITE
20,000°C (36,000°F)

PALE BLUE-WHITE
8,500°C (15,000°F)

3 There are many DIFFERENT TYPES OF STARS.

They range in brightness, size, colour, and behaviour.

A star's colour, temperature, and size change over the course of its lifetime.
WHITE
6,500°C (11,700°F)

YELLOW-WHITE
5,300°C (9,570°F)

4 A star's COLOUR tells us HOW HOT ITS SURFACE IS.

ORANGE
4,000°C (7,230°F)

5 There are approximately 200 BILLION TRILLION STARS.

Red stars are coolest
RED
3,000°C (5,430°F)

When it reaches the surface, lava cools and hardens quickly into an igneous rock.

1 ROCKS ARE CONSTANTLY BEING RECYCLED AND TRANSFORMED into different types of rock over millions of years. This is called the ROCK CYCLE.

2 There are three types of rock: IGNEOUS, SEDIMENTARY, and METAMORPHIC.

Rocks are broken up into small pieces by things such as wind, water, ice, plants, and animals. This is called weathering. Then the smaller rocks are moved from one place to another. This is called erosion (see p225).

3 IGNEOUS ROCKS form when molten rock called MAGMA cools. It may cool underground or after erupting from a volcano.

ERUPTING

WEATHERING AND EROSION

Rivers carry rock fragments (sediment) to the sea, where they settle on the seabed and become buried.

12
April
THE ROCK CYCLE

COOLING

RISES TO SURFACE OVER TIME

4 When SEDIMENTS (smaller rocks) are BURIED and COMPRESSED together, they form SEDIMENTARY ROCK.

COMPRESSION

HEAT AND PRESSURE

MAGMA

SEDIMENTARY ROCK

METAMORPHIC ROCK

MELTING

Metamorphic rock may get so hot it melts to become magma, and the cycle begins again.

5 METAMORPHIC ROCK forms when rock is BURIED deep underground, and is exposed to very HIGH PRESSURE and TEMPERATURES.

1

The word laser is an acronym. It stands for "LIGHT AMPLIFICATION by STIMULATED EMISSION of RADIATION".

The light from a torch contains many colours (see p194) with different wavelengths.

A laser emits light that is just one colour and all the waves are lined up.

13
April
LASERS

2

A laser emits a very powerful, CONCENTRATED BEAM OF LIGHT in just one colour.

It is most commonly red or green.

5

SELF-DRIVING CARS use lasers TO NAVIGATE.

They send out beams in all directions, then measure the time the reflected light takes to return to work out where objects are.

4

The MOST POWERFUL LASER being developed will produce a beam one million, billion, billion times BRIGHTER THAN the most intense SUNLIGHT!

3

Lasers can be used to PERFORM SURGERY, CUT MATERIALS, and are in many common devices such as barcode scanners.

1 Earth formed when bits of ROCKY AND ICY DEBRIS orbiting the Sun began to CLUMP TOGETHER.

They were attracted by gravity and grew to become the huge bodies that we now know as Earth and the other planets in the Solar System.

Earth is known as the Blue Planet because of its vast oceans of water (see p134).

Earth's landmasses are still moving, as they are shifted by the movement of tectonic plates (see p72).

2 Earth is in the "GOLDILOCKS" ZONE. It is just the RIGHT DISTANCE from the Sun for WATER TO REMAIN A LIQUID – essential for life.

It is the only place in the Universe that we know for certain is home to living things.

5 Earth's WATER came from OUTER SPACE.

Scientists think it arrived on the comets and asteroids that collided with early Earth.

Earth's atmosphere (see p26) traps heat and protects us from small asteroids and radiation.

3 We know about Earth's STRUCTURE because of EARTHQUAKES.

Scientists noticed that different types of earthquake wave arrived at monitoring stations at different times. This meant the waves must have travelled through rock of differing densities.

4 Earth has to travel 940 MILLION KM (584 MILLION MILES) over 365.24 days to complete ONE ORBIT of the Sun.

15
April
WHALES

Whales must return to the surface regularly to breathe air through their blowholes.

Blue whales use baleen plates to filter tiny krill from the water they gulp.

5 ORCA are actually a type of DOLPHIN.
Due to their size, they have been given the name "killer whales".

Whales have thick fat layers to survive in icy cold oceans.

1 There are TWO TYPES of WHALE: ones with teeth and ones without!
Toothed whales catch their prey with sharp teeth. Baleen whales sieve tiny creatures from the sea using comblike plates.

4 Whale POO is an incredible ocean FERTILIZER!
It fertilizes microscopic algae called phytoplankton and transfers nutrients from the deep ocean back to the surface.

2 A blue whale's HEART is the size of a GOLF BUGGY.
It has to pump 10 tonnes (9.8 tons) of blood around the whale's body and can be heard 3 km (2 miles) away.

3 Some HUMPBACK whales don't EAT for six months.
They live off fat reserves while they migrate thousands of kilometres to their breeding grounds.

16
April
ANIMAL SILK

New Zealand glow worms light up to attract prey to fly into their sticky traps.

5 SPIDERS will sometimes EAT THEIR SILK webs to get the most out of the ENERGY and PROTEIN they contain.

4 WEAVER ANTS produce silk to join together leaves to BUILD their NESTS.

A silkworm's cocoon is made of a single, continuous strand of silk.

1 New Zealand GLOW-WORMS secrete SILK from their MOUTHS.
They hang the sticky threads from cave ceilings to catch prey.

2 SPIDERS can produce up to SEVEN DIFFERENT TYPES of silk for different uses, from WEB-MAKING to WRAPPING.

3 SILKWORMS spin silk COCOONS to protect themselves when they undergo METAMORPHOSIS (see p227).

1 Earth's continents and the rock beneath the oceans form GIANT SLABS called TECTONIC PLATES (see p72).

The plates float like rafts on the hot, semi-liquid mantle below, which moves them slowly across the surface of the planet.

Pangaea

250 MILLION YEARS AGO

Laurasia

Gondwana

200 MILLION YEARS AGO

2 250 million years ago, all the planet's land was JOINED TOGETHER in one SUPER-CONTINENT called PANGAEA.

Since then, the plates have moved around until they became the seven continents that we know today.

3 The CRUST UNDER THE OCEANS IS MUCH DENSER than the crust that makes up the CONTINENTS (land).

17
April
MOVING CONTINENTS

4

Oceanic CRUST is constantly RECYCLED back into the MANTLE.

It is rare for oceanic crust to be more than 200 million years old.

5

INDIA was once JOINED to AFRICA.

80 million years ago, it broke away and drifted north, before crashing into Asia.

66 MILLION YEARS AGO

The Atlantic Ocean is growing wider by 2–5 cm (0.8–2 in) every year.

In places, it is possible to dive between the Pacific and Eurasian plates.

PRESENT DAY

As the plates move apart, a rift opens up and new oceanic crust is formed.

18
April
STAGGERING SOUNDS

KRAKATOA ERUPTION 310 DB

TUNGUSKA EVENT 300 DB

1 The **LOUDEST POSSIBLE SOUND** in air is 194 decibels.

Above that, the sound waves become shockwaves.

2 The eruption of **THE KRAKATOA VOLCANO IN 1883** is estimated to be the **LOUDEST SOUND** in history.

TSAR BOMBA 224 DB

SATURN V ROCKET 204 DB

3 In 1908, an **ASTEROID EXPLODED** in the sky above Tunguska, Russia, flattening 80 **MILLION TREES**.

4 The **LOUDEST-EVER MAN-MADE SOUND** was the Tsar Bomba – a **NUCLEAR BOMB** tested by the Soviet Union in 1961.

5 NASA's *Saturn V* rocket launches were **SO LOUD** that observers had to stand **SEVERAL KILOMETRES AWAY**.

THUNDER 120 DB

19
April
HOW THE MOON FORMED

1 The early Solar System was a place of **CHAOS**.

Among the rocky debris were a Mars-sized planet called Theia and a larger planet called Gaia.

2 4.5 billion years ago, the two **PLANETS COLLIDED**.

A cloud of gas, dust, and rock fragments sprayed into space.

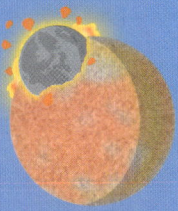

3 The **CORES** of Theia and Gaia **MERGED**, and Gaia became **EARTH**.

The remaining debris was pulled into a ring by Earth's gravity.

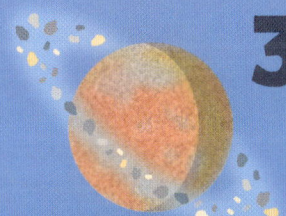

4 Over millions of years, the **DEBRIS COLLIDED AND CLUMPED** together to form the **MOON**.

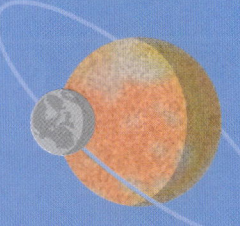

5 This is just one **THEORY** – the Moon may have formed from a series of **SMALLER COLLISIONS** instead of one strike.

20
April
CARBON CYCLE

Naturally, Earth's atmosphere contains about 0.4 per cent CO_2.

Some of the Sun's heat is trapped by CO_2 in the atmosphere, keeping the planet warm.

2
WITHOUT CARBON, Earth would be FROZEN AND LIFELESS.

Carbon dioxide (CO_2) – a gas in our atmosphere – acts like a giant blanket around the planet. Without it, Earth would be too cold for life.

1
CARBON (see p61) is ALWAYS ON THE MOVE.

On Earth, it travels between living things, our atmosphere, the oceans and the soil, in a process called the carbon cycle.

Plants use sunlight and CO_2 to make their food.

Animals, including humans, breathe out CO_2.

CO_2 is released into the atmosphere by burning fossil fuels in cars and factories.

Animals take in CO_2 when they eat plants.

Animals release CO_2 when they poo or die.

Oceans take in CO_2.

Dead sea life releases CO_2 or becomes fossil fuel (oil).

CO_2 is released from fossil fuels like coal and oil when they are dug up and burned.

When plants die, they rot or turn into coal (see p212).

3
Human activities are DISRUPTING THIS BALANCED SYSTEM.

Fossil fuels used in our homes, workplaces, and cars, as well as intensive farming, have upset the natural carbon cycle.

4
PLANTS TAKE IN MORE carbon dioxide THAN THEY GIVE OUT.

This is why forests are so vital in the carbon cycle.

5
CARBON DIOXIDE in the sea ALLOWS MARINE CREATURES TO MAKE HARD SHELLS of calcium carbonate crystals.

21
April
ARACHNIDS

1 There are about 65,000 SPECIES of ARACHNID.

The arachnid group includes spiders, scorpions, ticks, mites, and harvestmen.

2 VERY FEW arachnids CAN EAT SOLID PREY.

Instead, most of them liquify their prey and slurp them up like soup!

3 Unlike other arachnids, SCORPION MOTHERS GIVE BIRTH TO LIVE YOUNG.

The babies climb onto her back and she carries them around.

Arachnids have a protective outer skeleton called an exoskeleton.

4 The WORLD'S LARGEST SPIDER, the goliath bird-eating spider, is 30 cm (12 in) WIDE!

Arachnids have two body segments, with all the legs attached to the front segment.

5 The average person's bed contains MILLIONS OF DUST MITES. They're so small you usually can't see them.

All arachnids have eight legs.

22
April
STOMACH

1 Your stomach CAN STRETCH TO 75 TIMES ITS SIZE when full.

An empty stomach is the size of your fist, but folds in its walls means it can easily stretch and shrink back again.

2 It needs PROTECTING FROM ITSELF!

Gastric acid made by your stomach to digest food, could digest the stomach itself, so it secretes a sticky mucus barrier.

Three layers of muscle rhythmically squeeze food, turning it into a gloopy soup called chyme.

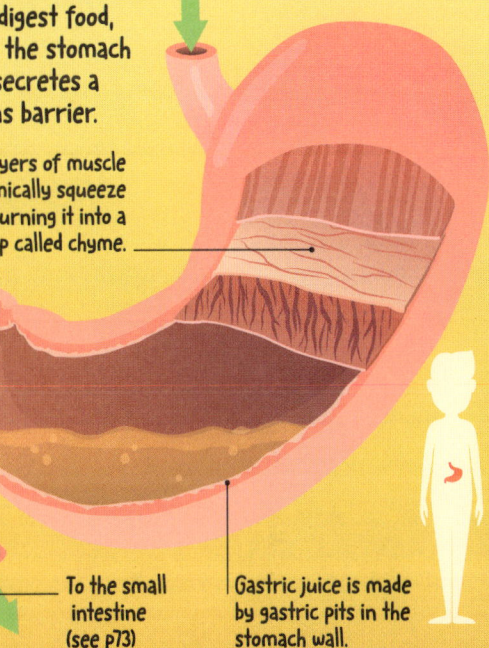

Food from the mouth

To the small intestine (see p73)

Gastric juice is made by gastric pits in the stomach wall.

3 Your stomach produces 3 litres (5.3 pt) of GASTRIC JUICE EVERY DAY.

4 Your STOMACH LINING is REPLACED EVERY FEW DAYS – to make sure that gastric juices can't leak out.

5 A RUMBLING STOMACH DOESN'T MEAN YOU'RE HUNGRY!

The rumbling sound comes from stomach muscles contracting and relaxing, which they do all the time. When it's empty, there's just no food to muffle it.

87

23 April
CONSTELLATIONS

1
Thousands of years ago, people would look for PATTERNS IN THE STARS and make up STORIES about the figures they found.

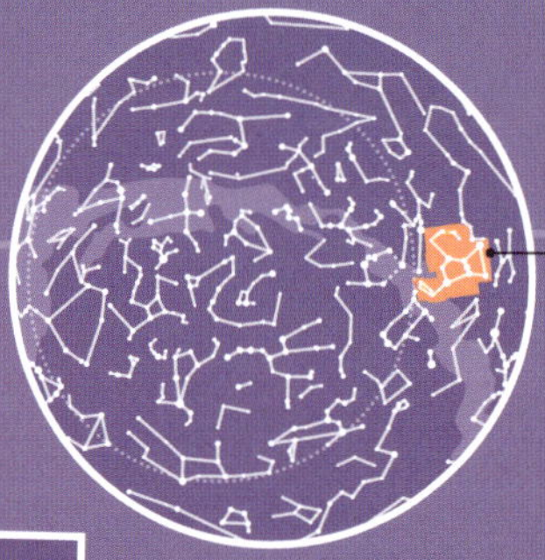

Orion is just one of 88 recognized constellations. They fit together like jigsaw pieces to form a map of the night sky.

2
In one area of sky, astronomers saw the shape of a HUNTER from Greek mythology called ORION.

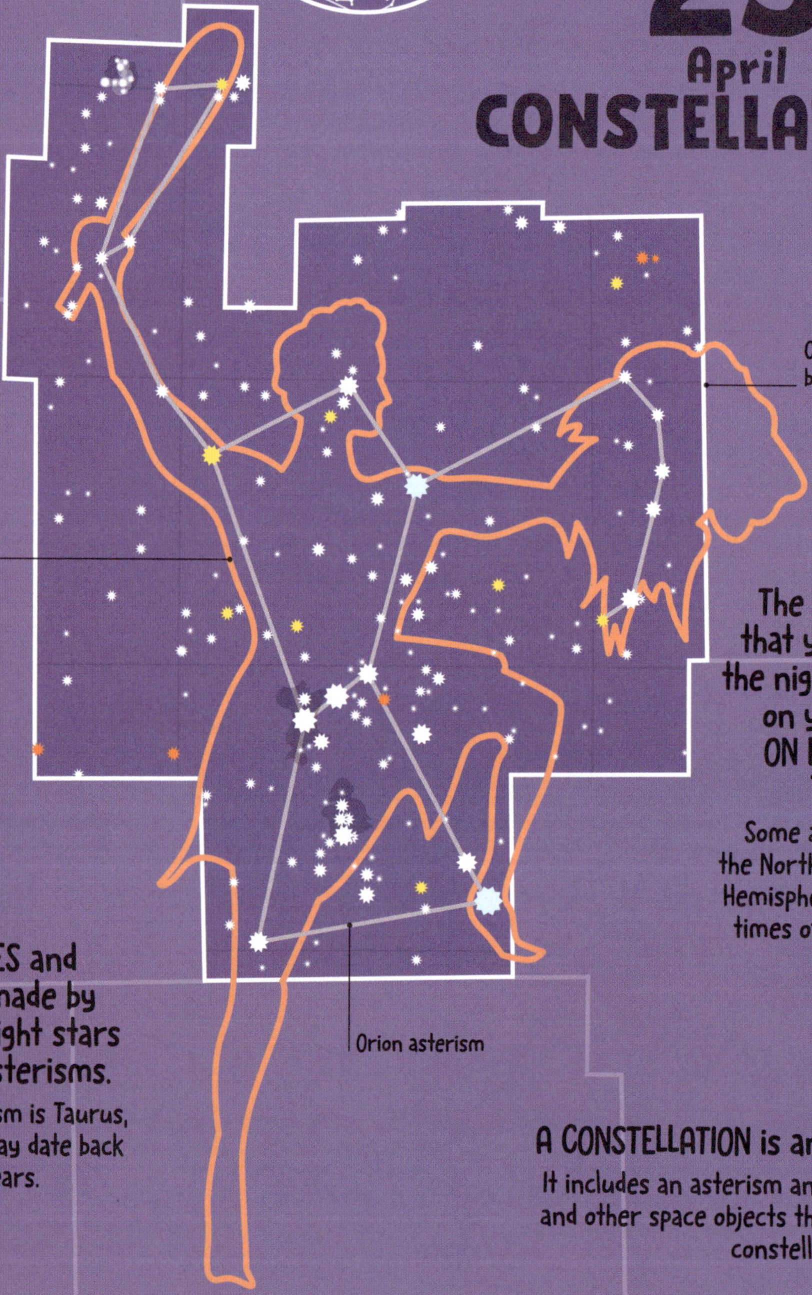

Many constellations represent legendary stories and myths. In Greek mythology, Orion the hunter was placed among the stars by Zeus.
Constellation boundary
Orion asterism

3
The SHAPES and PATTERNS made by joining up bright stars are called asterisms.

The oldest asterism is Taurus, the bull, which may date back 17,000 years.

4
The constellations that you can see in the night sky depend on your LOCATION ON EARTH and the TIME OF YEAR.

Some are only visible in the Northern or Southern Hemisphere, or at certain times of the year due to Earth's rotation.

5
A CONSTELLATION is an AREA of sky.

It includes an asterism and all of the stars and other space objects that fall within the constellation's boundary.

Batteries

1 Batteries are STORES of CHEMICAL ENERGY that can be converted to ELECTRICAL ENERGY (see p128).

2 To release energy, batteries need to be CONNECTED to an ELECTRICAL CIRCUIT (see p25).

3 The FIRST BATTERY was made of lots of DISCS OF COPPER AND ZINC stacked together.
It was invented in 1800 by the Italian physicist Alessandro Volta.

4 RECHARGEABLE batteries can be used up to 1,000 TIMES!
They are recharged by electricity.

5 The BIGGEST BATTERY SYSTEM in the world stores energy generated by SOLAR POWER.
It uses more than 120,000 batteries to store the energy.

24
April
BATTERIES

25
April
METEORS

Meteoroids are usually about the size of a pebble, but can be as small as a speck of dust.

1 METEORS are SHOOTING STARS – streaks of light that sometimes appear across the night sky.

2 Meteors occur when TINY FRAGMENTS of rock from asteroids and comets, called METEOROIDS, enter Earth's ATMOSPHERE.

3 As the meteor SLAMS INTO THE ATMOSPHERE, it heats up and GLOWS BRIGHTLY.

A meteor that appears brighter than Venus is called a fireball.

4 MILLIONS of meteoroids burn up in Earth's atmosphere EVERY DAY.

5 Any material that makes it to the GROUND is called a METEORITE.

The largest known intact meteorite was 2.7 m (9 ft) wide.

1 Plant cells are similar to animal cells, but have a more DEFINED SHAPE and several EXTRA ORGANELLES (see p16).

2 Each cell has a TOUGH CELL WALL and a large, round fluid store called A VACUOLE.

The wall prevents the cell from bursting when the vacuole expands.

3 Tiny GREEN structures called CHLOROPLASTS allow plants to get energy from sunlight.

They use this energy for photosynthesis (see p136).

The nucleus contains all of the cell's genetic information.

Under the cell wall, a thin membrane controls what substances enter and leave the cell.

The cell wall is rigid and supports the rest of the cell.

Water is stored in the large vacuole at the centre of the cell.

Green chloroplasts give plant leaves their colour.

Structures called mitochrondria generate energy for the cell.

A fluid called cytoplasm fills the inside of the cell.

26 April PLANT CELLS

4 When plant cells don't get enough water, they LOSE THEIR STRUCTURE and become floppy as THE VACUOLE SHRINKS.

This is what makes a plant appear to be wilting.

5 There are SPECIAL TYPES of plant cells.

Cells in the hairs of the roots (see p103) have long, pointy bits that allow them to absorb nutrients from the soil more easily. Cells in the stem are also adapted to transport water.

27 April
METEOR SHOWERS

1 Meteor showers occur when EARTH PASSES THROUGH A COMET'S DUST TAIL (see p42).

2 The dusty DEBRIS BURNS and GLOWS as it enters Earth's atmosphere.

This is seen as bright streaks in the sky called shooting stars or meteors (see p89).

3 Meteor SHOWERS HAPPEN AT ROUGHLY THE SAME TIME AND AREA of the sky each year.

This is because Earth passes through the orbits of comets at predictable points in the year.

4 The showers are usually NAMED AFTER A STAR or constellation.

The Perseids are named after the constellation Perseus, which the shower appears to radiate from.

5 During the peak of the GEMINID METEOR SHOWERS each December, as many as 100 METEORS CAN BE SEEN EVERY HOUR.

The bright streaks of the Geminid Meteor shower can be seen above the Very Large Array telescope in New Mexico.

28 April
FORMULA 1™

1 Formula 1™ (F1™) racing cars can REACH SPEEDS of 375 km/h (233 mph).

An air inlet above the driver sucks in air to cool the powerful engine.

2 F1™ car engines are FIVE TIMES MORE POWERFUL than regular car engines.

Front and rear wings increase downforce to help the car stick to the road and corner faster.

Front wing

3 EIGHT different TYPES OF TYRES are used on F1™ cars.

Smoother tyres are used in dry weather with those that offer more grip used in wet conditions

4 The car's shape is specially DESIGNED TO MAKE AIR FLOW AROUND IT smoothly.

This reduces air resistance, helping it to go faster.

5 F1™ cars produce so much DOWNFORCE that once at speed, they'd be able to DRIVE UPSIDE DOWN sticking to the ceiling.

29
April
TSUNAMIS

1

Tsunamis are HUGE, DESTRUCTIVE SURGES OF WATER caused by seabed LANDSLIDES, EARTHQUAKES, and VOLCANIC ERUPTIONS.

The sudden and violent seafloor movement displaces large amounts of water above it.

2

The TALLEST TSUNAMI WAVE measured over 500 m (1,700 ft).

It was recorded in 1958 when a tsunami was triggered by a landslide into a narrow bay on Alaska's coast.

5. The wave hits the shore with great force and can surge inland by several kilometres.

3. A tsunami wave forms at the surface and the energy travels out in all directions.

2. The seismic waves radiate out and push seawater upwards.

4. Water is drawn away from the shore, adding to the large wave.

SEISMIC WAVES

1. An underwater earthquake shifts the seabed, releasing huge amounts of energy in the form of seismic waves.

OCEAN FLOOR

OCEAN FLOOR

A detector on the seabed transmits data to a buoy at the surface.

5

The PACIFIC TSUNAMI WARNING SYSTEM is based in Hawaii, and uses a NETWORK OF DETECTORS TO MONITOR FOR QUAKES that might cause tsunamis, providing EARLY WARNING for those who might be affected.

3

On average, there are only TWO TSUNAMIS EACH YEAR.

4

Tsunamis can move AS FAST AS 970 km/h (600 mph) across the open ocean.

That's almost as fast as a jet plane!

1 FERMENTATION is a CHEMICAL REACTION involving MICROORGANISMS, such as bacteria and fungi.

2 Microorganisms CONSUME SUGARS IN A FOOD and CONVERT THEM into substances that add flavour and air.

3 CHEESE, BREAD, AND CHOCOLATE, as well as many other foods and drinks, are all MADE WITH THE HELP OF FERMENTATION.

4 The BLUE PARTS OF BLUE CHEESE ARE A FUNGUS CALLED MOULD (see p73)! It helps to ferment the cheese.

5 YEAST IS A TYPE OF FUNGUS that makes BREAD DOUGH RISE.

Yeast produces carbon dioxide gas as it consumes sugars, giving the bread an airy structure.

Over time, the yeast produces more and more carbon dioxide bubbles, so the dough rises.

1
May
MICROCHIPS

1 MICROCHIPS are TINY PIECES OF TECHNOLOGY found in computers, phones, and other electronic devices.

2 They are made of the ELEMENT SILICON (see p252) and contain SMALL ELECTRICAL CIRCUITS.

3 The FIRST MICROCHIPS were invented in the 1950s.

They made computers smaller and more efficient.

4 PETS are sometimes fitted with MICROCHIPS UNDER THEIR SKIN.

The chips store a unique code, so the pet can be identified if lost.

5 In 2022, technology company IBM created the SMALLEST, MOST POWERFUL MICROCHIP.

It is just 2 nanometres wide – too small to be seen with the naked eye.

2 May BEARS

1 There are EIGHT SPECIES OF BEAR.

These are: Asiatic black bears, American black bears, brown bears (which include grizzly bears), giant pandas, polar bears, sloth bears, spectacled bears, and sun bears.

2 GRIZZLY BEARS CAN EAT FOR 20 HOURS IN A SINGLE DAY.

They do it in preparation for hibernating, when they fast.

3 Hibernating bears BREATHE just ONCE OR TWICE A MINUTE.

Their body temperature drops and heart rate slows, too.

4 A NEWBORN GIANT PANDA IS REALLY TINY.

Just one-nine-hundredth the size of its mother!

5 On average, the POLAR BEAR is the LARGEST BEAR SPECIES.

A fully grown male can grow to a length of 3 m (10 ft).

The polar bear has two layers of fur and a thick layer of fat to keep warm.

Brown bears are very powerful with large heads in proportion to their bodies.

| SUN BEAR | SLOTH BEAR | AMERICAN BLACK BEAR | ASIATIC BLACK BEAR | GIANT PANDA | ANDEAN / SPECTACLED BEAR | BROWN / GRIZZLY BEAR | POLAR BEAR |

3 May ARTIFICIAL INTELLIGENCE

1 Artificial intelligence (AI) is technology that ALLOWS COMPUTERS TO LEARN and MAKE DECISIONS like humans.

2 YOU MAY HAVE USED AI.

When you ask your mobile phone a question, AI is used to recognize and analyse your words and respond.

3 A computer can use AI to play COMPLEX GAMES LIKE CHESS.

It makes choices based on a set of rules, and the more it plays, the better it gets.

4 SELF-DRIVING CARS USE AI.

They "see" the road and make decisions to avoid obstacles.

5 Some ROBOTS use a type of AI CALLED MACHINE LEARNING.

This allows them to recognize patterns, learn, and improve their performance at certain tasks.

1 GPS (Global Positioning System) is A SATELLITE NAVIGATION SYSTEM.

It identifies where people are on Earth.

Each satellite sends a unique signal and its position in orbit.

The satellites circle Earth twice a day in a precise orbit.

Mobile phone with map navigation app

4
May
GPS

2 TWENTY-FOUR SATELLITES (see p116) MAKE UP THE GPS SYSTEM.

They circle Earth in different orbits to cover all of its surface area.

3 TO CALCULATE WHERE A PERSON IS ON EARTH, the GPS receiver combines the positions of at least three satellites with how long their signals took to arrive.

The satellite's location and time the signal is sent is picked up by the map application on Earth.

The satellites are powered by solar energy.

5

GPS receivers can be used to STUDY THE EFFECTS OF EARTHQUAKES, by measuring HOW MUCH THE GROUND HAS MOVED!

They are also used by planes, ships, cars, and mobile phone apps for navigation.

4 Using MULTIPLE SATELLITES gives A MORE ACCURATE POSITION.

This process is called trilateration.

One satellite signal places the receiver somewhere at the edge of a sphere.

Edge of sphere

Using two satellite signals narrows the location down to where the two spheres intersect.

Intersection points show two possible locations

Adding a third satellite signal gives a precise position.

Location of GPS receiver

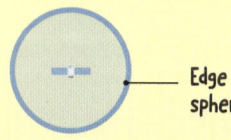

5 May
JELLYFISH

1 Jellyfish HAVE NO BONES, HEART, OR BRAIN.

Despite their name, jellyfish are not fish. They are invertebrates (see p253), with simple jelly-like bodies.

2 A jellyfish will often DRIFT THROUGH THE OCEAN, RIDING THE CURRENTS.

It can also move itself along by forcing jets of water out of its bell (main body).

Stinging cells along the tentacles help many jellyfish to hunt and defend themselves.

The bell of the mauve stinger jellyfish is just 10 cm (4 in) across, but its tentacles can reach 3 m (10 ft).

3 Most jellyfish begin life as A TINY LARVA, WHICH ATTACHES ITSELF TO A ROCK.

It goes through several life stages there, before floating free in its adult "medusa" form.

4 THE IMMORTAL JELLYFISH can live its life cycle IN BOTH DIRECTIONS.

It can change from its adult form back to its juvenile "polyp" form, so it might be able to live forever!

5 Some jellyfish species HAVE DEADLY STINGS.

The most dangerous in the world is the Australian box jellyfish, which can kill a human with a single sting.

6 May
RAINFORESTS

1 Scientists think that OVER HALF OF ALL PLANT AND ANIMAL SPECIES IN THE WORLD live in rainforest habitats.

2 The AMAZON is the LARGEST rainforest.

It covers around 6.7 million sq km (2.6 million sq miles) and is home to over 2.5 million species of insect alone!

3 MORE THAN 40 DIFFERENT species of ANT were found on A SINGLE TREE in Peru!

4 Nearly 80 PER CENT of all flowers in Australian rainforests are found NOWHERE ELSEWHERE ON EARTH.

5 There are probably LOTS OF PLANTS and ANIMALS living in rainforests that SCIENTISTS HAVEN'T DISCOVERED YET.

About 80 per cent of all creatures in rainforests can be found in the canopy layer of the trees.

EMERGENT LAYER

CANOPY LAYER

UNDERSTOREY LAYER

1 The Moon is 4.5 billion years old. Most scientists think that ANOTHER SPACE BODY COLLIDED WITH EARTH, SENDING DEBRIS INTO SPACE.

This debris was pulled into Earth's orbit, and eventually merged to form a sphere – our Moon.

2 WE always SEE THE SAME SIDE OF THE MOON from EARTH.

It takes the same time for the Moon to complete one spin on its axis as it does to orbit Earth, so the same side is always facing towards us.

As it orbits Earth, the Moon is slowly spinning.

THE EARTH

The distance varies because its orbit (path) around Earth is elliptical (a squashed circle).

The Moon has many craters, formed by asteroids bombarding it, especially in its early life.

3 THE MOON has 20 SEAS, 14 BAYS, 20 LAKES, and ONE OCEAN, but they're made of rock, not water!

Early astronomers thought the Moon's craters they saw were dried-up seas, so gave them names like "Sea of Tranquility".

4 EARTHQUAKES HAPPEN ON THE MOON.

The tremors, known as moonquakes, are usually weaker than on Earth, but can last much longer, up to 10 minutes.

7
May
THE MOON

5 On average, THE MOON IS 382,500 km (237,674 miles) AWAY FROM EARTH.

That's equivalent to 30 Earth-widths.

1

MAMMALS first appeared on Earth 225 MILLION YEARS AGO.

Early mammals were tiny shrew-like creatures. Mammals became bigger only after most of the dinosaurs were wiped out.

Brasilodon is one of the earliest known mammals. It was just 20 cm (8 in) long, and lived alongside early dinosaurs.

2

Mammals are WARM-BLOODED, have a BACKBONE, feed their young MILK, and have FUR OR HAIR.

8
May
MAMMALS

A long pregnancy allows an elephant's very complex brain to develop.

3

There are three types of mammal – MONOTREMES, MARSUPIALS, and PLACENTALS – based on how their young are born.

Placental infants, such as these baby mice, develop inside the mother before being born.

Baby kangaroo, also called a joey

Marsupial infants are born early and the baby climbs up into the mother's pouch to finish maturing.

Monotremes, such as the duck-billed platypus, lay eggs.

4

The female African ELEPHANT is PREGNANT FOR 22 MONTHS.

This is the longest pregnancy of all mammals. The shortest is just 12 days!

5

BATS are the ONLY TYPE of mammal that can FLY.

A bat's wings are made of a skin-like membrane.

The membrane is supported by finger-like bones.

9 May
THE LIFE OF OUR SUN

1 THE SUN STARTED LIFE as a new star, or PROTOSTAR.

Gravity squashed the material in its core, making it very hot. Eventually, nuclear reactions began in its core and the young Sun started to glow.

2 How a star AGES depends on its MASS.

The more massive the star, the faster it uses up its fuel and the shorter it lives. Our Sun is a medium-mass star.

3 Stars like the Sun GLOW for BILLIONS OF YEARS.

As they run low on fuel, they swell into dim stars known as red giants.

4 Eventually, the Sun will SHED ITS OUTER LAYERS.

They will form a glowing cloud of material called a planetary nebula. At the centre will be a small, hot core called a white dwarf.

5 This is JUST ONE WAY a star can EVOLVE.

Low-mass stars age slowly until they shrink into black dwarves, while high-mass stars expand and explode as supernovas (see p24).

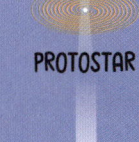
PROTOSTAR

YOUNG SUN

THE SUN TODAY

RED GIANT

PLANETARY NEBULA

WHITE DWARF

---

10 May
FIBRE OPTICS

1 FIBRE OPTICS is a way of SENDING DATA in the form of LIGHT.

2 It works by CONVERTING DATA from a computer into LIGHT SIGNALS.

The light signals are sent along strands of glass called optical fibres.

Light reflects off fibre walls

3 OPTICAL FIBRES rely on REFLECTION.

The beams of light bounce down the length of the fibre by reflecting off the internal surface of the glass.

Bundle of fibres

4 HUNDREDS of fibres can be BUNDLED TOGETHER into CABLES.

Cables like these are used for phones, television, and the internet.

Cable jacket

5 Fibre optic cables are MORE EFFICIENT THAN COPPER WIRES.

Light travels very fast, so huge amounts of data can be sent quickly, and little of the energy is lost as heat.

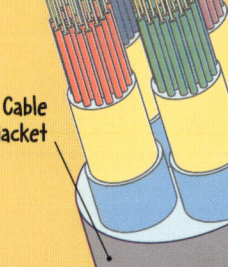

1

A dwarf planet is a SPACE OBJECT that ORBITS THE SUN, is LARGE ENOUGH TO BE MOSTLY ROUND, and has NOT CLEARED ITS PATH of other orbiting objects.

Makemake is just one-ninth of the width of Earth.

2

There are FIVE dwarf planets in our Solar System: CERES, PLUTO, HAUMEA, MAKEMAKE, and ERIS.

But there may be more still to be discovered.

Dysnomia is the only moon of Eris.

Eris is so far from the Sun it takes 557 Earth years to make one orbit.

3

All five dwarf planets are SMALLER THAN EARTH'S MOON, but some of them have MOONS OF THEIR OWN.

Ceres is found in the inner Solar System, in the asteroid belt between Mars and Jupiter.

Haumea spins so fast that one day lasts just four hours.

Namaka, one of Haumea's moons.

4

HAUMEA is an EGG-SHAPED dwarf planet. It is ONE OF THE FASTEST-SPINNING OBJECTS in the Solar System.

Hi'iaka, a moon of Haumea.

11
May
DWARF PLANETS

Pluto shares its orbit with other objects, so it cannot be classified as a planet. It has five moons.

5

PLUTO used to be called a PLANET.

However, in 2006 scientists decided Pluto didn't meet the criteria of a planet and created the new category of dwarf planet.

12 May
OXYGEN

OTHER
NITROGEN
HYDROGEN
CARBON
OXYGEN

1 Oxygen is the MOST COMMON ELEMENT on Earth's SURFACE.

It is a gas, but combines with other substances to make water and many of Earth's minerals.

2 It makes up about 21 PER CENT OF EARTH'S ATMOSPHERE.

We depend on the oxygen in the air to stay alive.

3 Oxygen makes up TWO-THIRDS OF THE HUMAN BODY.

It is found mostly in the form of water.

4 Oxygen REACTS VERY EASILY with OTHER SUBSTANCES.

Rust forms when iron or steel reacts with the oxygen in the air.

5 STARS PRODUCE OXYGEN deep inside their cores in NUCLEAR FUSION REACTIONS.

13 May
LIFE BEYOND EARTH

1 DO ALIENS EXIST? Maybe, but astronomers CAN'T YET BE SURE!

2 FOR LIFE TO EXIST ON ANOTHER PLANET, the planet must lie in a "GOLDILOCKS ZONE": it must orbit the RIGHT DISTANCE from a star for its temperature to SUPPORT LIFE.

3 Astronomers have confirmed MORE THAN 5,600 PLANETS beyond the Solar System.

However, only a tiny fraction of these lie in a Goldilocks Zone.

4 Scientists at the Search for Extraterrestrial Intelligence (SETI) Institute SEARCH FOR ALIEN LIFE.

They use radio telescopes to search for artificial radio signals that indicate life.

5 There are probably BILLIONS OF HABITABLE PLANETS in the Universe.

But it may be that life is so rare that we are alone!

14
May
SKYSCRAPERS

1
BUILDING SKYSCRAPERS is an ENGINEERING CHALLENGE.

This is because they have to withstand their own weight, the force of heavy winds, and sometimes even earthquakes.

2
Skyscrapers only BECAME POSSIBLE after the INVENTION OF THE LIFT, in 1853.

No one would want to regularly walk up hundreds of flights of stairs to get to work!

Its shape is designed to break up the flow of wind to stop the building from swaying.

3
The TALLEST SKYSCRAPER IN THE WORLD is the BURJ KHALIFA, in Dubai.

At 828 m (2,717 ft) high, it is nearly three times the height of France's Eiffel Tower.

4
In countries that have EARTHQUAKES, skyscrapers are built on a layer of material that ABSORBS THE ENERGY from a quake. Some also contain HUGE WEIGHTS THAT SWING AS THE BUILDING MOVES, to reduce its sway.

The weight swings the opposite way to the building, pulling it back to the centre.

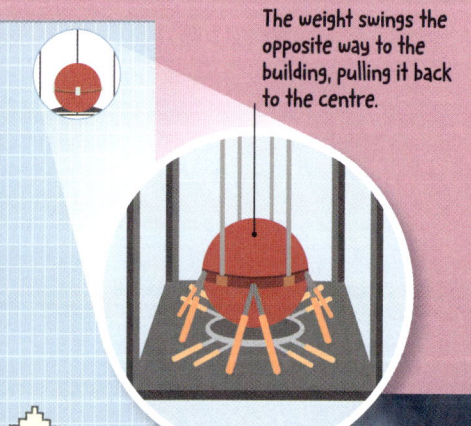

The Shanghai Tower has 128 floors above ground and another five below ground.

5
The 632-m (2,073-ft) tall SHANGHAI TOWER in China has a TWISTED SHAPE TO WITHSTAND HIGH WINDS.

It also has 947 steel tubes drilled 60 m (200 ft) into the ground.

1 Usually extending deep underground, ROOTS ANCHOR PLANTS TO THE GROUND, and EXTRACT NUTRIENTS FROM THE SOIL.

15 May
ROOTS

Some plants, such as grasses, have a network of shallow, fibrous roots.

Most plants have one large central root called a taproot.

2 Roots can be VERY STRONG – BREAKING THROUGH ROCKS AND EVEN CONCRETE as they grow!

3 TINY HAIRS COVER EACH ROOT, made up of a UNIQUE TYPE OF CELL (see p90).

The hairs give the roots a larger surface area for absorbing water.

4 The roots of the WINTER RYE PLANT can STRETCH 623 km (387 miles)!

Extensive root systems can connect multiple trees or plants together.

5 Some plants have ROOTS ABOVE GROUND.

These "aerial" roots can help climbing plants cling to walls or trees or give roots that extend into water a bit of air.

103

1 A WETLAND is a place where the LAND IS COVERED BY WATER for some or all the time.

Even when not covered by water, the soil in wetlands stays soggy and waterlogged.

More than 650 bird species make their home in the Pantanal wetlands, including the hyacinth macaw.

16 May WETLANDS

2 Wetlands can develop in MOST CLIMATES, and are found on EVERY CONTINENT EXCEPT ANTARCTICA.

The water that covers them can be saltwater, freshwater, or brackish (a mixture of both).

Wetland plants filter harmful bacteria and toxins out of the water through their roots.

Jaguars will hunt almost anything, from a crocodile to a capybara.

South America's Pantanal wetland is the largest in the world.

3 40 PER CENT of the world's plants and animals DEPEND ON WETLANDS.

However, wetlands are disappearing three times faster than forests and 25 per cent of wetland species are facing extinction.

The Pantanal has the highest concentration of crocodiles in the world.

The lily pad of the giant water lily can stretch up to 3 m (9.8 ft) across.

The roseate spoonbill swings its bill from side to side under water to filter out shrimp and insects.

The capybara, the world's biggest rodent, is the size of a labrador dog.

4 GREEN ANACONDAS, the world's HEAVIEST SNAKES, live on land and in water.

The heaviest anaconda ever found was 227 kg (500 lb) and more than 8 m (27 ft) long.

5 A SHOAL OF PIRANHAS COULD EAT A CAPYBARA to the bone IN ONE MINUTE.

But these small predators usually feed on much smaller prey, including other piranhas!

1 Plasma is ELECTRICALLY CHARGED GAS. It is known as the fourth state of matter (see p134).

2 MOST OF THE MATTER IN THE UNIVERSE is made of plasma.
It is what stars are made from.

3 In plasma, some negatively charged ELECTRONS are SEPARATED FROM THEIR ATOMS (see p53), which causes the plasma to have an electrical charge.

4 Plasma is usually VERY HOT. THIS HEAT IS WHAT CAUSES ITS ATOMS TO SEPARATE.
Plasma can be hotter than the Sun!

An electrode at the centre of a plasma ball sends out an electric current that flows out through the plasma and creates colourful, dancing tendrils.

5 We can create plasma BY RUNNING ELECTRICITY THROUGH A GAS, such as neon, creating a colourful plasma ball.

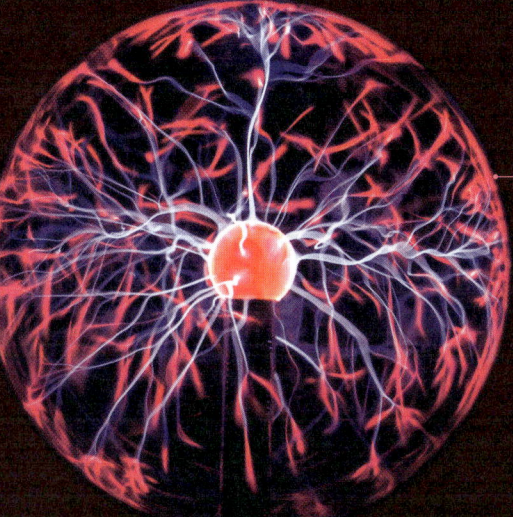

---

1 SLIME MOULDS are not moulds or fungi but simple, single-celled organisms called PROTISTS.

2 Most are found in DAMP, SHADY HABITATS, such as ROTTING WOOD AND LEAF LITTER, where they engulf bacteria and fungi.

3 Slime moulds move around to SEARCH FOR FOOD.
Some can travel at speeds of up to 1.35 mm (0.05 in) per second.

4 *Physarum polycephalum* has NO BRAIN, but can SEARCH ITS WAY THROUGH A MAZE to find food.

5 There are more than 1,000 SPECIES of slime mould, with names such as SCRAMBLED EGG and DOG'S SICK SLIME.

Physarum polycephalum sends out branches to find food.

It retracts itself from unsuccessful paths.

Once food is found, it directs its growth towards the food.

It lays a chemical trail to "remember" where it has been.

105

The hard rock is more resistant to erosion, so it overhangs.

1

Most waterfalls form because SOFT ROCK erodes (wears away) quicker than HARD ROCK.

They are usually carved out of the land by powerful river waters.

The soft rock is easily worn away, so the water gradually carves it out below the hard rock.

3

The area at the BOTTOM OF A WATERFALL is called a PLUNGE POOL.

The plunge pool grows when rocks topple in and get churned around by the water.

2

Waterfalls form where a RIVER flows over a layer of HARD ROCK onto a layer of SOFT ROCK.

Over time, the soft rock erodes, leaving an overhanging wall of hard rock behind, and the river water plunges over the ledge.

Fallen rocks are thrown around by the falling water, which wears away more rock and creates a plunge pool.

4

Waterfalls can also form ALONG FAULTS (cracks) in EARTH'S CRUST.

Blocks of rock can rise or fall, creating rock walls for rivers to flow over.

The force of the falling water makes the water in the plunge pool swirl around.

5

The world's TALLEST WATERFALL is ANGEL FALLS in Venezuela. The water plunges 979 m (3,212 ft).

20
May
VOYAGER MISSIONS

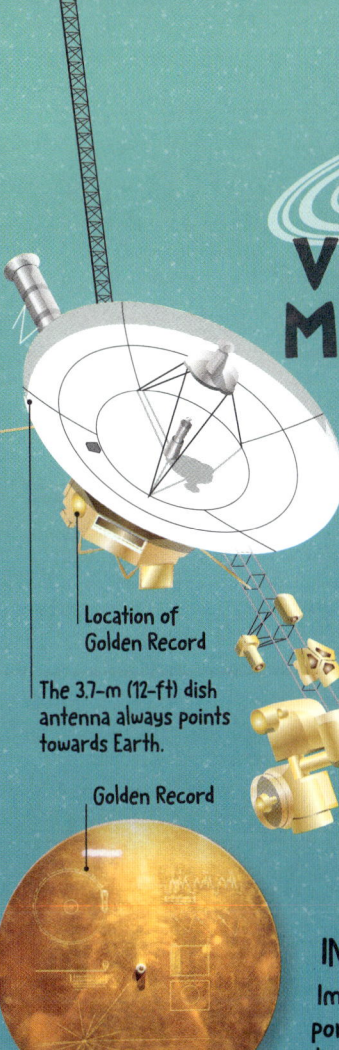

Location of Golden Record

The 3.7-m (12-ft) dish antenna always points towards Earth.

Golden Record

1 The two VOYAGER SPACECRAFT were launched in 1977. The original mission to explore Jupiter and Saturn was designed to last just five years.

2 They carried MESSAGES in case they were INTERCEPTED BY ALIENS. Images and sound recordings portraying life on Earth were stored on a gold-plated copper disc called the Golden Record.

3 After flying by Jupiter and Saturn, they visited URANUS and NEPTUNE, and SENT BACK PICTURES as they flew past.

4 They have now become the FIRST CRAFT to LEAVE OUR SOLAR SYSTEM, and are still sending back data.

5 Voyager 1 is now the FURTHEST HUMAN-MADE OBJECT from Earth. It is more than 24 billion km (15 billion miles) away.

21
May
RADIO WAVES

1 Like microwaves and X-rays, RADIO WAVES are a type of radiation. They are an invisible wavelength of light on the electromagnetic spectrum (see p204).

2 Radio waves have WAVELENGTHS (see p217) that vary from 1mm (0.04 in) to HUNDREDS OF KILOMETRES long.

3 Unlike the visible light we can see, radio waves PASS RIGHT THROUGH OBSTACLES like walls, trees, and even the human body.

4 ASTRONOMERS use radio telescopes to see the STARS AND GALAXIES HIDDEN behind DUST CLOUDS deep in the Universe.

5 Without radio waves, the INTERNET WOULDN'T WORK. Phones, TVs, laptops, satellites, Bluetooth earbuds, and Wi-Fi routers all rely on radio waves to SEND AND RECEIVE DATA.

1

The liver is the body's LARGEST INTERNAL ORGAN.

An adult liver is about the size of a football. It plays an important role in digestion and cleaning the blood.

Oxygen-rich blood comes from the heart to give the liver energy.

Clean, low-oxygen blood returns to the heart.

2

The liver FILTERS around 1.4 LITRES (2.5 pt) OF BLOOD EVERY MINUTE.

Nutrient-rich blood from the intestines flows through the liver, which processes it and stores or releases minerals, vitamins, iron, and glucose as needed by the body.

Right lobe

Left lobe

3

It is made up of 100,000 TINY PROCESSING UNITS called LOBULES.

Each one is shaped like a hexagon and is the size of a sesame seed.

4

Your liver is A MANUFACTURING PLANT.

Among other things, it makes hormones (see p257); proteins to build tissue; and bile, a fluid that helps digest fat.

LIVER

STOMACH

Processed blood flows out of the lobule through a central vein.

The hepatic portal vein brings nutrient-rich blood from the small intestine.

Gallblader stores bile

INTESTINES

Cells inside the lobule extract nutrients and produce bile.

22
May
LIVER

5 YOUR LIVER CAN REGROW!

Unlike most parts of the body, you could lose up to 75 per cent of your liver and it would regrow completely within a few months.

Nutrient-rich blood flows into the lobule.

23
May
HYDROPOWER

1 Hydropower (or hydroelectric power) uses THE FLOW OF WATER TO GENERATE ENERGY.

2 Humans have harnessed this by BUILDING LARGE DAMS that store water in reservoirs above a slope.

Dam wall

Power lines carry electricity away.

Dam gates open and water rushes through.

Turbines convert movement energy into electricity.

3 Opening the dam gates releases a POWERFUL FLOOD OF WATER DOWNHILL, WHICH FLOWS OVER TURBINES.

4 The THREE GORGES DAM in China can GENERATE 20 TIMES THE ENERGY of a single nuclear power station.

5 The HOOVER DAM on the Arizona-Nevada border, USA, is as tall as a 60-STOREY BUILDING.

Hoover Dam

24
May
GLUE

1 The earliest known glue was used 200,000 YEARS AGO. It was made of tar from THE BARK OF BIRCH TREES.

2 The SECRET of SUPER GLUE is WATER! When the glue leaves its tube, it reacts with water vapour in the air, which makes it harden almost instantly.

3 In 2019, the WORLD'S STRONGEST HUMAN-MADE GLUE was used to lift a 17.2-tonne (38,000-lb) truck into the air. The truck was suspended for an hour before the glue gave way.

4 White PVA glue works because the WATER IT CONTAINS EVAPORATES. The water leaves behind sticky polymers (see p127).

5 A bacterium called *Caulobacter crescentus* makes THE STICKIEST KNOWN NATURAL MATERIAL. It is three times stickier than super glue!

109

1 NITROGEN is ESSENTIAL FOR LIFE.
It is found in the proteins and DNA in all living things.

2 The air we breathe is 78 PER CENT NITROGEN.
But plants and animals can't absorb nitrogen straight from the air.

25
May
NITROGEN CYCLE

Lightning can turn nitrogen gas in the air into compounds that will dissolve in rain.

5 PLANTS ABSORB NITRATES through their roots and CONVERT THEM INTO PROTEIN.
Animals obtain these proteins from eating the plants, and use them to build the proteins they need.

Denitrifying bacteria release nitrogen back into the air.

When plants and animals die, they put nitrogen back into the soil.

Animal waste, such as poo, returns nitrogen to the soil.

SOIL NITRATES

4 BACTERIA in the soil can "FIX" NITROGEN.
They turn nitrogen in the air into nitrates, which can dissolve in water for plants to absorb.

Nitrifying bacteria that turn nitrogen into nitrates are an essential part of the nitrogen cycle.

3 Some BACTERIA, along with FUNGI, digest dead plant and animal matter to RELEASE NITROGEN INTO THE SOIL.

26 May
LIQUIDS

1

LIQUID is a STATE OF MATTER (see p134) between SOLID and GAS.

2

A liquid's PARTICLES ARE LOOSELY BONDED, so liquids can FLOW and CHANGE SHAPE to take on the shape of the container they are in.

3

VISCOSITY is a measure of HOW EASILY A LIQUID FLOWS.

Honey is very viscous, so it flows slowly, whereas water pours quickly and easily.

Lava is molten rock that comes from deep underground.

4

If you STIR SALT INTO WATER, it DISAPPEARS!

In fact, it just mixes with the water. This is called dissolving and the mixture it creates is called a solution.

5

Some liquids can be EXTREMELY HOT – LAVA can reach temperatures of MORE THAN 1,100°C (2,000°F).

---

27 May
SPACECRAFT

1 Anything designed to FLY THROUGH SPACE is a SPACECRAFT.

Satellites and the International Space Station are both types of spacecraft. Only a few spacecraft carry human crew.

2 UNCREWED spacecraft have VISITED EVERY PLANET in our Solar System.

3 NASA's ORION spacecraft will TAKE HUMANS BACK TO THE MOON.

In 2022, it orbited the Moon carrying crash test dummies to monitor the effects of radiation on future crew.

4 The FASTEST SPACECRAFT is the PARKER SOLAR PROBE.

It will reach speeds of more than 644,000 km/h (400,000 mph).

5 The EUROPA CLIPPER's solar arrays stretch LONGER THAN A BASKETBALL COURT.

It's the biggest interplanetary spacecraft ever built... so far!

NASA's Europa Clipper will visit Europa (moon of Jupiter) to see if it could support life.

28 May
FAR SIDE OF THE MOON

1 WE ALWAYS SEE THE SAME SIDE OF THE MOON from Earth because of the SPEED OF ITS SPIN.

2 The FAR SIDE OF THE MOON WASN'T SEEN UNTIL 1959 when a Russian space probe sent photos back to Earth.

There are fewer dark seas of cooled lava.

The far side is densely covered with impact craters.

3 The far side has COUNTLESS MORE IMPACT CRATERS. Scientists think its thicker crust makes it trickier for lava to emerge and cover them.

4 Humans HAVEN'T STEPPED ON THE FAR SIDE of the Moon.

China's Chang'e 4 lander

5 Only two spacecraft have landed there... so far. China's Chang'e 4 and Chang'e 6 missions landed on the far side in 2019 and 2024.

29 May
RIVERS

1 The NILE RIVER in Africa is the LONGEST RIVER IN THE WORLD. It winds for 6,650 km (4,132 miles).

2 Most rivers BEGIN IN MOUNTAINS OR HILLS. The start of a river is called the source.

Small streams that feed into a larger river are called tributaries.

3 Rivers wear away land over THOUSANDS TO MILLIONS OF YEARS to form VALLEYS.

4 BENDS in a river are called MEANDERS. If they grow too close together, they form an oxbow lake.

5 When rivers meet the sea, they can DEPOSIT SEDIMENT, forming a FAN-SHAPED LANDFORM CALLED A DELTA.

A river's water usually comes from springs or snow melt in the mountains.

Steep-sided valleys are called gorges.

An oxbow lake forms when a meander is cut off.

Where a river meets the sea is called the mouth.

1

PRESSURE is the SIZE OF A FORCE on a certain area.

Solids, liquids, and gases can all exert pressure. Putting pressure on an object can make it move or change shape.

Pushing on a balloon changes its shape.

2

A force can produce HIGH OR LOW PRESSURE, depending on the SIZE OF THE AREA IT PUSHES ON.

The smaller the area a force acts on, the greater the pressure it creates.

A finger doesn't pop a balloon because its tip has a large area.

A pin pops a balloon because its tip has a small area.

3

AIR MOLECULES ARE ALWAYS MOVING and bouncing off the walls of their container.

The force they exert on the wall of their container is called air pressure.

A balloon stays inflated because of the pressure created by the air inside.

30
May
PRESSURE

5

At the summit of Mount Everest, in Asia, atmospheric pressure is one-third that of sea level.

ATMOSPHERIC PRESSURE is the pressure on EARTH'S SURFACE exerted by the WEIGHT OF AIR.

Atmospheric pressure is greatest near the ground (sea level) and falls as altitude increases.

Air particles are squashed together by the weight of particles above.

4

WATER exerts PRESSURE too.

At the bottom of the Mariana Trench in the Pacific Ocean, the pressure is more than a thousand times greater than at the surface. Only a few tough life forms can withstand the pressure.

The weight of the water overhead in the Mariana Trench is equal to 1,800 adult elephants!

1

Your immune system is YOUR DEFENCE AGAINST DISEASE.

One way your body defends itself is with white blood cells that hunt and destroy germs in your body that could make you unwell.

2

There are TWO MAIN TYPES OF WHITE BLOOD CELL in your body: macrophages and lymphocytes.

They are constantly being made in your bones and released into the bloodstream.

Cytoplasm stretches around bacteria.

MACROPHAGE

Cell nucleus (control centre)

3 LYMPHOCYTES attack pathogens (germs) by PRODUCING PROTEINS CALLED ANTIBODIES.

The antibodies stick to the germ and immobilize it for a macrophage to engulf and destroy it.

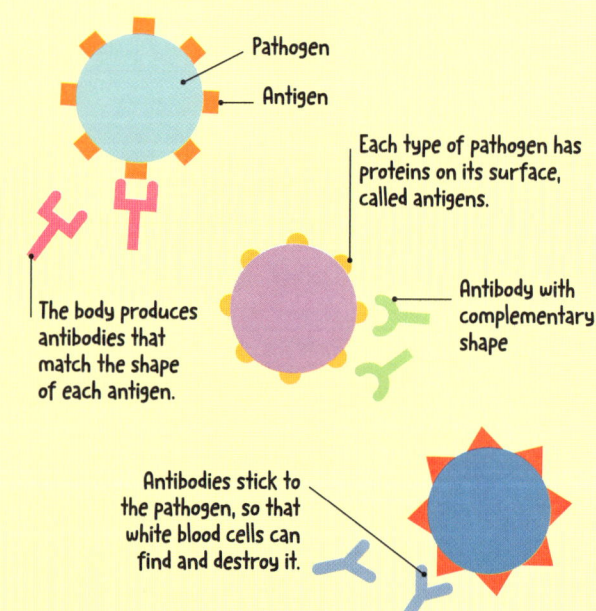

Pathogen

Antigen

Each type of pathogen has proteins on its surface, called antigens.

The body produces antibodies that match the shape of each antigen.

Antibody with complementary shape

Antibodies stick to the pathogen, so that white blood cells can find and destroy it.

4 MACROPHAGES patrol your body in your bloodstream, SWALLOWING UP and DIGESTING ANYTHING FOREIGN THAT THEY FIND.

This could be dirt from a cut, a virus, or bacteria.

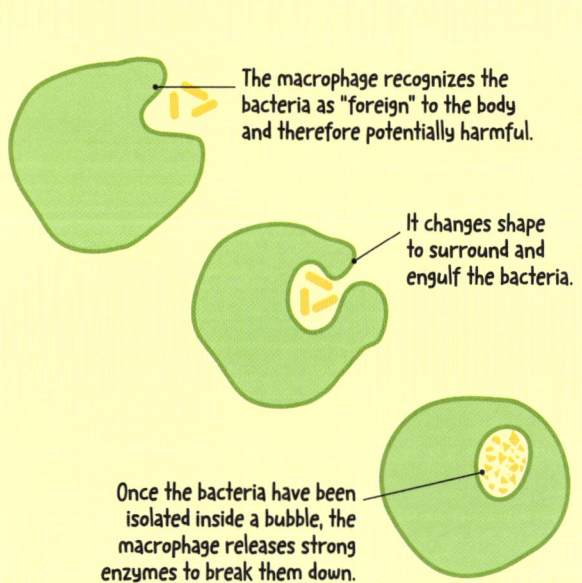

The macrophage recognizes the bacteria as "foreign" to the body and therefore potentially harmful.

It changes shape to surround and engulf the bacteria.

Once the bacteria have been isolated inside a bubble, the macrophage releases strong enzymes to break them down.

5 After you have EXPERIENCED a GERM, you have GREATER IMMUNITY to it.

Special memory cells remember the germ, and act faster to neutralize it next time.

1 June
VACCINES

1
Vaccines work by **TEACHING YOUR BODY** to recognize and defeat germs **WITHOUT HAVING CAUGHT THEM BEFORE.**

This is especially important if the diseases they cause are life-threatening.

2
Most vaccines contain **ALTERED, WEAKENED, or DEAD GERMS.**

These may give you mild symptoms, but do not put you at risk of harm.

3
THE BODY REMEMBERS the (harmless) germ in case it **MEETS IT AGAIN.**

As with natural immunity, memory cells are primed by the vaccine, and ready to make antibodies really quickly.

4
The **FIRST VACCINE** was popularized in 1796 by **DR EDWARD JENNER.**

It protected people from smallpox, which killed millions of people every year.

5
The COVID-19 VACCINE made the **BODY'S OWN CELLS MIMIC THE VIRUS.**

The genetic code from the virus was used to make the body's muscle cells produce antigens (virus proteins), so the body knew what to fight.

Vaccine particle

White blood cell

Matching antibody

Antibodies are released and spread through the body, to stick to more germs.

Macrophage cells swallow anything the antibodies stick to.

The original cell also produces memory cells.

Memory cells stay in the bloodstream for years and will remember the virus if it returns.

1 A satellite is AN OBJECT IN SPACE that orbits around a bigger object.

There are many human-made, or artificial, satellites orbiting Earth.

2 Sputnik 1 was THE FIRST SATELLITE LAUNCHED INTO SPACE in 1957. It was made by the USSR (Russia).

2
June
SATELLITES

Satellites can be launched into different orbits at different distances from Earth.

3 Satellites can TRACK THE WEATHER, take photos, and MAKE COMMUNICATION BY PHONES AND THE INTERNET POSSIBLE.

4 CUBESATS are less than 10 cm (4 in) across.

They are microsatellites that could fit in the palm of your hand!

5 There are more than 10,000 ACTIVE SATELLITES in orbit!

There are also many more disused ones that add to Earth's space junk (see p233).

---

1 The FIRST WHEELS were made THOUSANDS OF YEARS AGO!

They were used on roads around 3500 BCE in ancient Mesopotamia (modern Iraq).

2 Wheels were FIRST USED TO SPIN LUMPS OF CLAY to make pottery jars and bowls.

3
June
WHEELS

3 Alloys such as STEEL (see p221) are the most COMMONLY USED MATERIALS to make wheels today.

The oldest wheels were made from wood.

4 The LONGEST CAR RECORDED HAD 26 WHEELS, and one enormous mining vehicle has a massive 760 wheels!

Wheels can be made lighter by using spokes instead of solid wood or metal.

5 LEGO MAKES MORE TYRES for the wheels of its toy cars than any other tyre manufacturer!

It makes more than 300 million a year.

1

Reptiles are COLD-BLOODED VERTEBRATES with scaly skin.

They live on land and in water, and most species lay eggs.

Snakes (see p15), such as this creamsicle corn snake, use their muscles to climb trees.

2

The main groups of reptiles are LIZARDS AND SNAKES, CROCODILIANS, and TURTLES AND TORTOISES.

There are more than 10,000 individual species.

Like other lizards, chameleons are covered in tiny scales.

4
June
REPTILES

3

Tortoises and some turtles CAN RETREAT THEIR HEAD AND LIMBS entirely into their shell when threatened.

African spurred tortoise

Its shell actually makes up part of its skeleton.

5

Spiny tuataras are the RAREST TYPE OF REPTILE – found ONLY IN NEW ZEALAND.

They have survived since the age of the dinosaurs.

4

The SALTWATER CROCODILE is the BIGGEST REPTILE.

It can grow to lengths of 7 m (23 ft) and has a ferociously strong bite.

"Salties", as they are called in Australia, are ambush hunters like most other crocs.

117

5
June
MARS

Deimos, which means "dread"

Polar ice cap

Phobos, which means "fear"

1 Mars is the FOURTH PLANET from the Sun and our planetary NEIGHBOUR.

Venus is actually nearer to us, but totally inhospitable for visits.

Mars has lots of iron in its soil, which rusts and turns the surface and atmosphere red.

Olympus Mons

Ascraeus Mons

Pavonis Mons

Arsia Mons

Carbon dioxide is the main gas in Mars' thin atmosphere. Strong winds cause dust storms on the surface.

2

Mars may look hot but, being further from the Sun than Earth is, IT'S PRETTY CHILLY.

The average temperature is about –65°C (–85°F).

Valles Marineris

3

MARS HAS TWO MOONS: PHOBOS AND DEIMOS – the twin sons of the Greek god of war, Ares (who the Romans called Mars).

4

You'd be able to JUMP MUCH HIGHER on Mars because it has ONE-THIRD OF THE GRAVITY ON EARTH.

You'd be able to jump three times higher, to about the height of a standard door!

5 A day on Mars is about the same as on Earth, but A MARTIAN YEAR IS 687 EARTH DAYS LONG – nearly two Earth years!

6
June
SONAR

Waves are sent out from a device under a ship.

5 DOLPHINS USE A NATURAL FORM OF SONAR known as echolocation (see p130) to navigate and detect prey.

The time between the waves being sent out and coming back reveals how far away objects are.

1 SONAR is a sensing system that USES SOUND WAVES TO "SEE" UNDERWATER.

4 Some fishing boats use sonar to FIND SHOALS OF FISH.

2 Sonar was INVENTED IN 1912 to help ships AVOID ICEBERGS hidden under the water.

The waves travel through water but bounce back off any solid objects.

3 Sonar has allowed us to MAP THE OCEAN FLOOR.
It scans the floor in sections to map every bump and dip.

7
June
BIRDS' EGGS

Two membranes protect the egg from bacteria.

The chick will have eaten the yolk by the time it hatches.

The shell has tiny holes for oxygen to enter.

3 The YOLK is the CHICK'S FOOD.
It supplies the nutrients the chick needs to allow it to grow.

Chalaza

4 The ALBUMEN (egg white) is like a WRAPAROUND CUSHION for the egg yolk and growing chick.
It is used as food for the chick once the yolk is used up.

1 An EGG contains EVERYTHING A CHICK NEEDS TO DEVELOP as it grows from a cluster of cells into a young animal.

Yolk

Chalaza

5 Two cords called CHALAZAE STOP THE YOLK FROM MOVING around the egg.
They attach the yolk to both ends of the egg.

2 The shell is INCREDIBLY STRONG for its thickness.
The dome shape spreads any external pressure all over the egg's structure.

Albumen

Air sac

119

8
June
GETTING INTO SPACE

1 Scientists haven't agreed on WHERE SPACE BEGINS, but many put it at 100 km (60 miles) ABOVE SEA LEVEL.

2 To OVERCOME the strong pull of EARTH'S GRAVITY (see p75), you need a rocket that can travel at least 11 km (7 miles) PER SECOND.

3 To ESCAPE EARTH'S ORBIT and go deeper into space, a spacecraft needs A BOOST OF SPEED AND THRUST.

4 It must FLY IN A CURVED PATH fast enough to FALL INTO EARTH'S ORBIT.
This ensures gravity won't pull it back to Earth.

5 TIMING IS ESSENTIAL: space scientists ensure Earth is NEAR THE PLANET OR MOON they want to send a spacecraft to!

9
June
SOLID OR LIQUID?

1 A FLUID that BEHAVES STRANGELY when a force is applied to it is called a NON-NEWTONIAN FLUID.

2 OOBLECK, a mixture of cornstarch and water, will RUN THROUGH YOUR FINGERS if you pick it up, but STAY SOLID if you run across its surface.

3 When a snail is still, its SLIME IS SOLID and KEEPS THE SNAIL STUCK to a wall.
When the snail moves, pressure on the slime turns it liquid again.

4 KETCHUP BEHAVES LIKE A SOLID until you SHAKE its bottle.
Shaking applies pressure to the ketchup, making it behave like a fluid.

5 FROGS have NON-NEWTONIAN SALIVA.
When a frog's tongue strikes prey, its saliva briefly turns runny and envelops the prey, then thickens when the pressure drops, so the prey sticks to its tongue.

A frog's saliva is a non-Newtonian fluid.

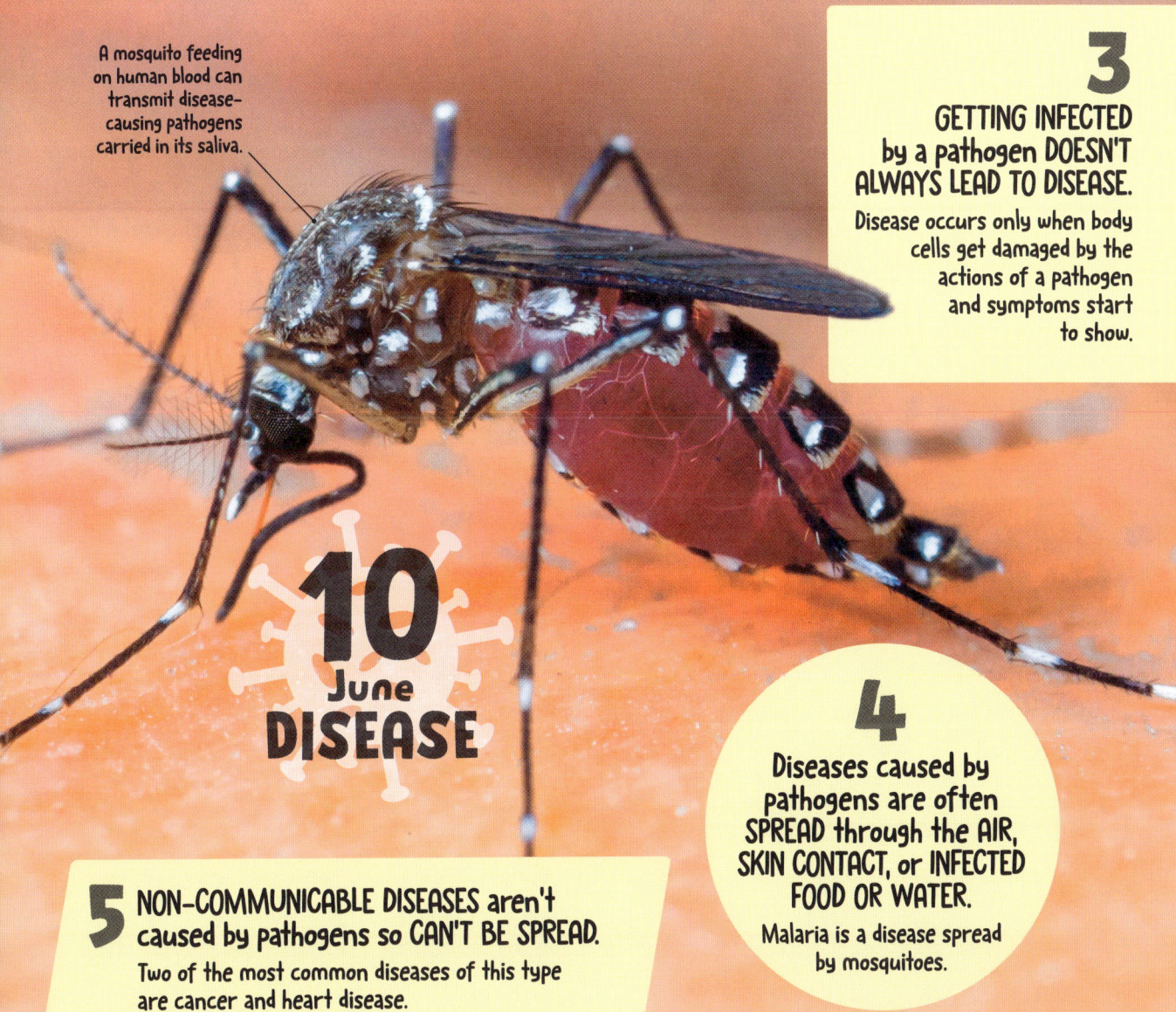

1

A DISEASE is a CONDITION that STOPS THE BODY FUNCTIONING PROPERLY or makes you feel unwell.

Diseases can affect one or multiple organs or body systems.

2

Diseases can be caused by GENETICS, a person's LIFESTYLE or the ENVIRONMENT they live in, AGEING, and HARMFUL PATHOGENS (disease-causing organisms) such as bacteria and viruses.

A mosquito feeding on human blood can transmit disease-causing pathogens carried in its saliva.

3

GETTING INFECTED by a pathogen DOESN'T ALWAYS LEAD TO DISEASE.

Disease occurs only when body cells get damaged by the actions of a pathogen and symptoms start to show.

10
June
DISEASE

4

Diseases caused by pathogens are often SPREAD through the AIR, SKIN CONTACT, or INFECTED FOOD OR WATER.

Malaria is a disease spread by mosquitoes.

5 NON-COMMUNICABLE DISEASES aren't caused by pathogens so CAN'T BE SPREAD.

Two of the most common diseases of this type are cancer and heart disease.

1

Blood has FOUR MAIN INGREDIENTS: plasma, red blood cells, white blood cells, and platelets.

Just over half of it is plasma, a watery, yellowish fluid. About 44 per cent is red blood cells that carry oxygen; 1 per cent is a mix of white blood cells that attack germs, and platelets that help to heal wounds.

Plasma

White blood cells and platelets

Red blood cells

2

IRON IS WHAT MAKES BLOOD RED.

Red blood cells are packed with a protein called haemoglobin. When they pick up oxygen in the lungs, iron in the haemoglobin reacts with it and turns red.

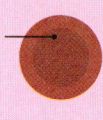 Oxygen-poor blood returning to the heart is much darker.

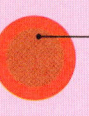 Oxygen-rich blood leaving the heart is bright red.

3

NEW RED AND WHITE BLOOD CELLS are made by bone marrow inside your bones (see p66).

The body makes 2 million new red blood cells every second!

The long bones in your body are the main factories for new blood cells.

4

A RED BLOOD CELL has a LIFESPAN of about 120 DAYS.

In that time, it travels about 480 km (300 miles), making 170,000 circuits around the body, passing through the heart, kidneys, and lungs.

White blood cells patrol your blood, attacking germs.

Red blood cells transport oxygen from the lungs to every cell in your body.

Artery

Plasma contains many things, such as hormones, proteins, and antibodies.

Platelets are small cell fragments that clump together to heal wounds.

5

At the end of their life, RED CELLS ARE EATEN BY OTHER BLOOD CELLS.

A macrophage (a special type of white blood cell) recognizes a worn-out red blood cell, engulfs it, breaks it down, and releases the iron into the blood to be used again.

11
June
BLOOD

12
June
FIBONACCI SEQUENCE

1 A sequence is a series of numbers or shapes THAT FOLLOW A PARTICULAR RULE.

2 In the Fibonacci sequence, EACH NUMBER in the sequence is THE SUM OF THE TWO PREVIOUS NUMBERS.

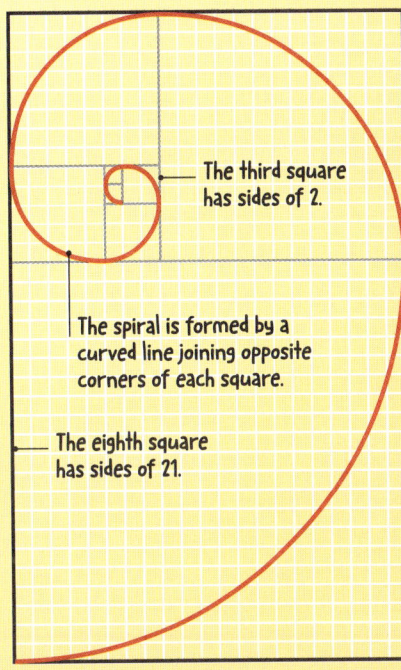

The third square has sides of 2.

The spiral is formed by a curved line joining opposite corners of each square.

The eighth square has sides of 21.

3 THE SEQUENCE GOES: 0, 1, 1, 2, 3, 5, 8, 13, 21, 34...

4 THE SEQUENCE CAN BE USED TO FORM A SPIRAL.

Imagine each number is a square with sides the length of that number. As the the squares get bigger you can arrange them to form a spiral, as on the left.

5 THE FIBONACCI SPIRAL can be FOUND THROUGHOUT NATURE and IN ARCHITECTURE.

The Fibonacci spiral can be seen in some seashells.

---

1 An asteroid is a SMALL, ROCKY OR METALLIC SPACE OBJECT that orbits the Sun.

2 Asteroids are leftovers from the BIRTH OF THE SOLAR SYSTEM.

13
June
ASTEROIDS

EARTH

Asteroid Belt

MERCURY

THE SUN

VENUS

3 THE MAJORITY OF ASTEROIDS are found in THE ASTEROID BELT.

This is a ring of asteroids orbiting the Sun between Mars and Jupiter.

MARS

4 The Asteroid Belt contains MILLIONS OF ASTEROIDS.

Over a million of them are more than 1km (0.6 miles) across.

5 AN ASTEROID PROBABLY WIPED OUT THE DINOSAURS.

It struck Earth about 66 million years ago and was the size of a small city.

JUPITER

1

Geothermal energy HARNESSES the ENERGY from HOT PARTS of Earth's CRUST (see p15).

2

Power stations PUMP COLD WATER UNDERGROUND, where it is HEATED up by hot rock or water.

The returned hot water is then used to power a turbine.

3

RESERVOIRS of hot water can be more than 2,000 m (6,500 ft) BELOW THE GROUND!

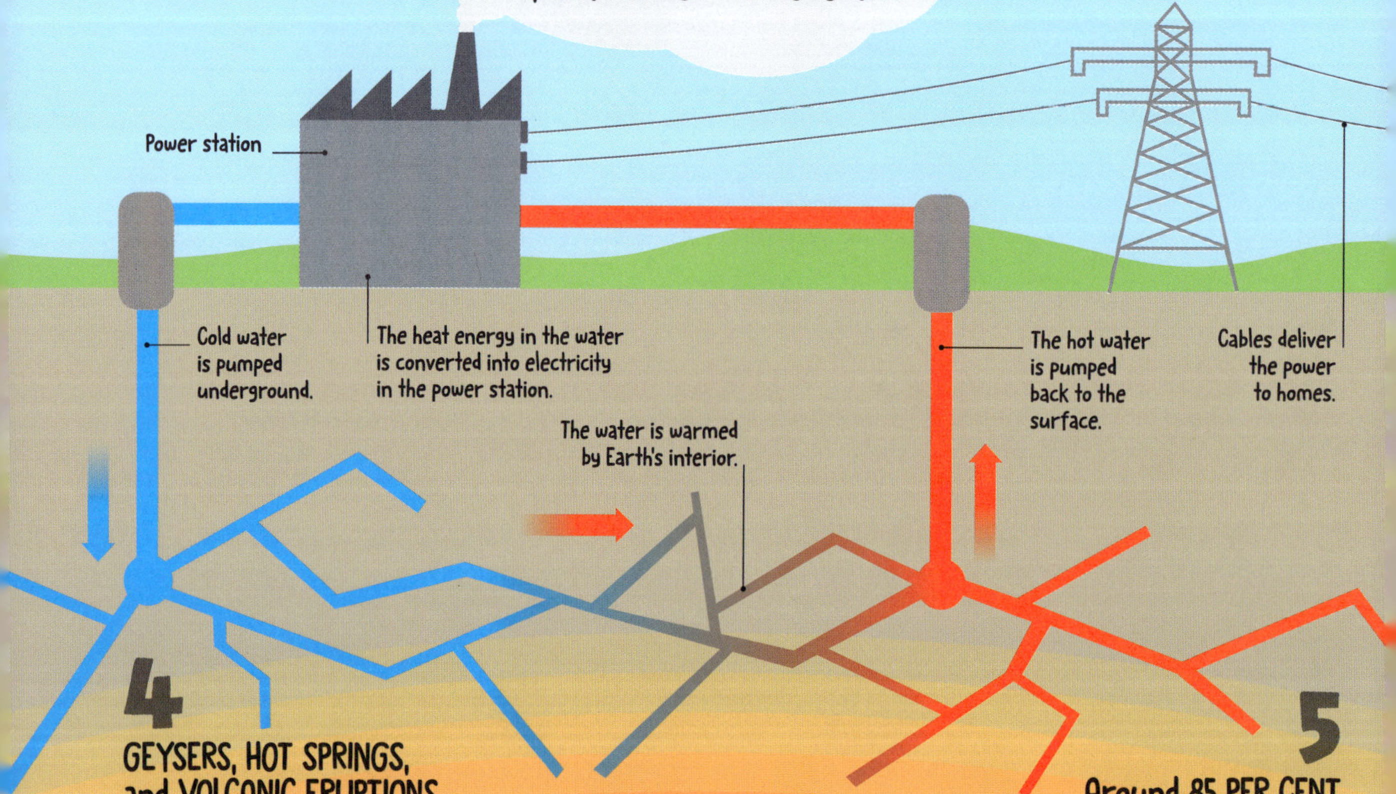

Power station

Cold water is pumped underground.

The heat energy in the water is converted into electricity in the power station.

The water is warmed by Earth's interior.

The hot water is pumped back to the surface.

Cables deliver the power to homes.

4

GEYSERS, HOT SPRINGS, and VOLCANIC ERUPTIONS are all NATURAL displays of geothermal energy.

5

Around 85 PER CENT of HOMES IN ICELAND are HEATED using geothermal energy.

The country sits on a rift between two tectonic plates, so has plenty of hot spots.

14
June
GEOTHERMAL ENERGY

124

15 June
MOBILE PHONES

1 Before mobile phones, people used SEMI-PORTABLE technology, like short-range WALKIE-TALKIES and large, bulky CAR PHONES.

Early "mobile" phones had big, heavy battery packs.

2 The first truly portable phone was the 1984 Motorola DynaTAC 8000X.

It weighed more than four times as much as today's phones and was known as the "brick phone".

3 Today's smartphones are like COMPUTERS.

A touchscreen detects and records your inputs, an antenna sends and receives information as radio waves, and a circuit board processes data.

Early smartphones could send emails and provided internet access.

4 Smartphones are 900 MILLION TIMES faster than the computers that guided Apollo 11 to the MOON!

5 There are now MORE MOBILE PHONES on the planet than PEOPLE!

More than 15 billion mobiles exist worldwide.

16 June
ALGAE

1 Algae are SIMPLE, PLANT-LIKE organisms that live in WATER or WET PLACES.

A teaspoon of seawater contains up to a million microscopic algae.

2 If you leave a GLASS OF WATER on a SUNNY WINDOWSILL for two weeks, algae will GROW in it and turn the water GREEN.

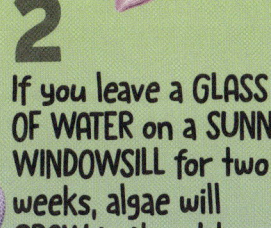

3 Algae use the ENERGY in SUNLIGHT to make FOOD.

This process, called photosynthesis (see p136), produces oxygen.

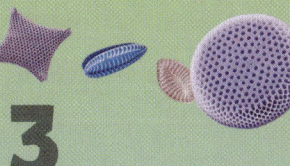

4 Without algae, we COULDN'T BREATHE.

At least half the oxygen in Earth's atmosphere comes from algae.

5 You brush your teeth with FOSSILIZED ALGAE! TOOTHPASTE contains ground-up fossils of algae called DIATOMS.

Their glassy shells make toothpaste gritty.

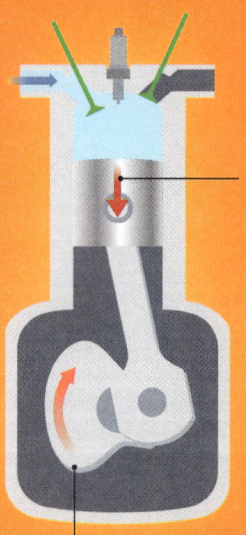

1. The piston moves down, pulling fuel and air into the cylinder.

A moving part called the crankshaft rotates, pulling down the piston.

1 Engines BURN FUEL to release heat.

This is then turned into movement energy and used to power a wide range of vehicles.

Spark plug

Fuel injectors

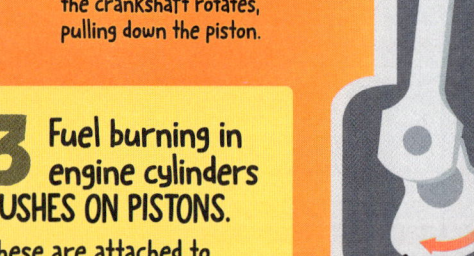

2. The piston moves up, squashing the fuel and air mix.

3. A spark ignites the fuel, which spreads out and pushes the piston down again.

2 Most cars (see p205) are POWERED BY INTERNAL COMBUSTION ENGINES, made up of lots of CYLINDERS.

Each cylinder goes through a four-stroke cycle as it burns fuel.

3 Fuel burning in engine cylinders PUSHES ON PISTONS.

These are attached to a connecting rod that transfers this movement energy to the wheels.

As the piston moves, the crankshaft turns.

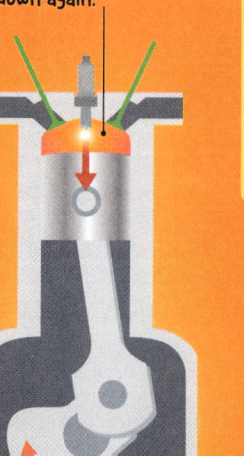

4. The piston moves back up, pushing out the waste gases, which are sent out of the car's exhaust pipe.

Jet engines are located below a plane's wings, so they can easily access air.

As the piston is forced back down, the crankshaft rotates. This movement is passed onto the wheels.

4 Planes usually have huge JET ENGINES.

These suck in air continually so they can supply lots of energy quickly.

5 The BIGGEST ENGINES are found inside giant container ships.

These can be taller than a house and weigh as much as 15 blue whales!

18 June RAIN

1

Rain is a TYPE OF PRECIPITATION (water that falls from clouds in the sky).

It is not the only type of precipitation: other examples include snow, sleet, and hail.

2

Rainfall is an incredibly important PART OF EARTH'S WATER CYCLE (see p154).

It fills lakes and rivers, recharges underground aquifers, and provides drinking water for plants and animals.

3

EARTH ISN'T THE ONLY PLACE where it RAINS.

Venus experiences extremely corrosive acid rain. On Saturn's largest moon, Titan, it rains methane, and on one exoplanet it rains glass!

4

Rain really DOES HAVE A SMELL.

It is called "petrichor", and is the smell of oils released by plants during dry weather to stop roots and seeds growing. As raindrops hit the ground, they throw these oils up into the air.

5

The WETTEST PLACE ON EARTH is Mawsynram in the Indian state of Meghalaya, which receives an average of 12 m (39 ft) of rain each year.

19 June POLYMERS

1 Polymers are everywhere - from NATURAL SILK, rubber, and DNA to SYNTHETIC PLASTICS such as PVC and polystyrene.

2 Polymers are LONG chain-like MOLECULES.

They form when many small molecules called monomers join together to form long chains.

3 SYNTHETIC POLYMERS, such as plastics (see p157), ARE POPULAR for their DURABILITY.

Polythene is formed by lots of ethene monomers. Their double bonds break and they form a chain of single bonds.

The atoms in this ethene monomer are joined by a double bond.

4 The PROPERTIES OF POLYMERS depend on how their molecules are bonded.

Polymers such as rubber are bendy and stretchy, but others such as PVC are rigid and hard.

5 Polymers can be BULLETPROOF.

The polymer Kevlar is five times stronger than steel and is used in bulletproof vests and spacecraft.

20
June
COMPUTERS

ENIAC took up a whole room when completed in 1946.

Computer engineers programmed ENIAC by adjusting cables and switches.

ENIAC

1 The first computers were MECHANICAL COUNTING MACHINES.

Charles Babbage, an English mathematician, designed one called the "difference engine" in the 1820s.

2 ENIAC, one of the EARLIEST electronic computers, had 6,000 SWITCHES!

3 ALL INFORMATION stored on a computer is converted into the NUMBERS 1 AND 0 – a system called BINARY.

4 The FIRST LAPTOPS were AS HEAVY AS FOUR BRICKS when they were introduced in the 1980s.

5 Today's SMARTPHONES are THOUSANDS OF TIMES MORE POWERFUL than the SUPERCOMPUTERS of the 1980s.

---

1 ENERGY is what MAKES THINGS HAPPEN. It powers EVERYTHING, from machines to the human body!

At the top of a slide, some energy is stored in you as gravitational potential energy.

The Sun's energy travels to Earth as light and heat energy.

5 Most of the ENERGY ON EARTH COMES FROM THE SUN.

Plants convert the Sun's heat and light energy into other forms.

Potential energy turns into kinetic energy when you move.

Some kinetic energy may be transferred to sound energy.

Chemical energy stored in food is transferred to you when you eat.

2 Energy CANNOT be CREATED OR DESTROYED.

It can only be transferred between different energy stores.

Springs store elastic potential energy when they are stretched or compressed.

3 There are many DIFFERENT WAYS energy can be TRANSFERRED, including through heat and sound.

4 BATTERIES are a STORE OF ENERGY.

Energy can be stored as well as transferred, such as in food or a coiled spring.

21
June
ENERGY

22
June
CIRCULATORY SYSTEM

5 CAPILLARIES are seriously small. They are just ONE-TENTH OF THE DIAMETER OF A HUMAN HAIR, and blood flows through them one cell at a time.

Capillaries link arteries and veins, and carry blood to individual cells.

The aorta is the body's largest artery.

4

If you joined them in a line, the body's BLOOD VESSELS would be LONG ENOUGH TO WRAP AROUND EARTH'S EQUATOR TWICE.

Veins take blood back to the heart. They have thin walls.

Heart

Arteries carry blood out from the heart. They have thick, muscular walls.

1

The CIRCULATORY SYSTEM is your body's BLOOD TRANSPORT NETWORK.

It has three parts: a complex web of blood vessels, the blood that flows through them, and your heart.

Upper body

Heart sends oxygen-rich blood around the body.

Right lung

Heart sends oxygen-poor blood to the lungs.

Lower body

Left lung

2

The circulatory system is both a DELIVERY AND RUBBISH COLLECTION SERVICE.

Blood brings essential nutrients and oxygen to the body's cells. It also carries away their waste.

3

The circulatory system is actually a DOUBLE SYSTEM.

Each beat of the heart sends blood to the lungs to pick up oxygen, while also sending oxygen-rich blood to the body's tissues.

23 June
ECHOLOCATION

1 Some animals, including BATS AND DOLPHINS, use ECHOLOCATION to AVOID OBJECTS and FIND FOOD.

Low-frequency echoes mean the moth is flying away.

High-frequency echoes mean the moth is flying towards the bat.

Sound waves sent out from bat

Large ears help bats hear the returning echoes.

2 They send out PULSES OF HIGH FREQUENCY SOUND WAVES.

Then they listen for the echoes created when the sound waves bounce off objects.

4 Some bats have ELABORATE NOSE "LEAVES" that FOCUS echolocation pulses.

3 The TIME IT TAKES for ECHOES TO RETURN tells a bat HOW FAR AWAY its prey is.

5 The FREQUENCY of returning waves also tells a bat whether its prey is FLYING AWAY OR TOWARDS THEM.

24 June
HYDROTHERMAL VENTS

1 Deep at the BOTTOM OF THE OCEAN are CRACKS IN THE SEAFLOOR where hot springs, known as HYDROTHERMAL VENTS, GUSH HOT WATER.

2 Hydrothermal vents are found where Earth's huge TECTONIC PLATES SPREAD APART (see p72).

3 The water spewed from the vents can be AS HOT AS 400°C (750°F).

But it doesn't boil, due to the extreme pressure at the depths where the vents are found.

The hot water picks up minerals along the way.

The water is heated by hot rock underground.

4 BACTERIA USE THE HEAT AND CHEMICALS from the vents to MAKE SUGARS (food).

This essential process, known as chemosynthesis, supports life around the vents.

Yeti crabs use their hairy claws to trap bacteria to eat.

5 A community of STRANGE ORGANISMS HAVE ADAPTED to thrive around the vents.

Vents are fed by cold water that seeps underground through tiny cracks in the seafloor.

25
June
POLLINATION

1

POLLINATION is part of HOW FLOWERS REPRODUCE.

It involves tiny grains called pollen being taken from plant to plant.

2

INSECTS like bees are ATTRACTED TO FLOWERS because they contain sugary NECTAR.

When they visit a flower to drink nectar, grains of pollen stick to their bodies.

A broad stigma catches pollen.

Dusty grains of pollen sit atop tall stalks called stamens.

A pollen tube grows down to the flower's ovaries.

3

When a pollen-coated insect VISITS ANOTHER FLOWER, the POLLEN LANDS ON ITS STIGMA.

A tube grows from the pollen to the flower's ovary and fertilizes an egg, which will later grow to become a seed.

A flower's tiny eggs are stored in ovules.

Insects seek out nectar from glands at the base of the stamen.

4

Around 90 PER CENT of FLOWERING PLANTS are POLLINATED BY ANIMALS.

By looking after pollinating animals like bees, we help many species of plants to survive too!

A single pollen grain seen with a powerful microscope

5 Individual POLLEN GRAINS measure around 25 microns - less than the WIDTH OF A HUMAN HAIR.

Each grain is a single cell. Their shapes and structures vary from plant to plant.

1 This COLOSSAL PLANET is the FIFTH from the Sun and the LARGEST in our Solar System.

In fact, Jupiter is so big that it has 2.5 times the mass (amount of matter) of all the other planets in the Solar System combined.

Io has hundreds of active volcanoes, making it the most volcanically active object in the Solar System.

2 Jupiter is a "GAS GIANT", which means it is made mainly of HYDROGEN gas and has NO SOLID SURFACE.

Jupiter's four largest moons (Ganymede, Callisto, Europa, and Io) are called the Galilean moons.

The Great Red Spot is more than 16,000 km (10,000 miles) wide: large enough to swallow up Earth.

Callisto has more craters than any other object in the Solar System.

The storm is at least 190 years old.

Europa has a saltwater ocean that scientists think may contain life.

3 Jupiter's "GREAT RED SPOT" is actually a storm.

It has been raging for hundreds of years, with wind speeds reaching 680 km/h (425 mph).

5 The coloured STRIPES around Jupiter are created by SWIRLING GAS IN ITS ATMOSPHERE.

As the gases move, different chemicals are pulled to the surface, creating the many colours that we see.

Jupiter's huge size means it has a lot of gravity, so it can pull space rocks into its orbit.

26
June
JUPITER

4 Jupiter has MORE THAN 90 KNOWN MOONS, and there are probably still MORE TO BE DISCOVERED.

Ganymede is the largest moon in the Solar System.

27
June
GOLD

1 GOLD CAME FROM OUTER SPACE 4.5 billion years ago when Earth formed, and sank deep inside Earth.

2 Some of the gold we mine from Earth's crust arrived later, when ASTEROIDS CARRYING GOLD BOMBARDED EARTH'S SURFACE.

3 YOUR BODY CONTAINS GOLD!

It is a tiny amount, weighing no more than a grain of sand. It would take gold from 30,000 humans to make a gold ring.

4 All the gold ever mined would FILL THREE OLYMPIC SWIMMING POOLS.

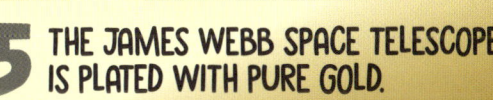

5 THE JAMES WEBB SPACE TELESCOPE IS PLATED WITH PURE GOLD.

Gold doesn't lose its shine and is also extremely good at reflecting infrared light, which the James Webb Space Telescope uses.

Each of the 18 sections has a thin layer of gold.

THE JAMES WEBB SPACE TELESCOPE

28
June
SHARKS

1 SHARKS WERE AROUND ON EARTH long before the DINOSAURS.

The earliest shark fossil ever discovered is from 420 million years ago.

2 The WHALE SHARK can grow to 18 m (60 ft) LONG, and yet it feeds on tiny plankton (see p255).

Under the snout are receptors that detect the electric pulses given off by a prey's muscles.

Nostrils on either side can smell a drop of blood in the water from great distances.

The lateral line down each side of the body picks up movement nearby.

GREAT WHITE SHARK

3 The SMALLEST SHARK is the dwarf lantern shark, WHICH IS ONLY AS LONG AS A PENCIL!

4 Most BABY SHARKS are BORN WITH TEETH... ready to hunt!

5 A SHARK'S LIVER STOPS IT FROM SINKING.

Sharks' livers are enormous – about 25–30 per cent of their body weight – and full of oil, which is lighter than water, so it keeps them afloat!

29
June
OCEANS

PACIFIC OCEAN

ATLANTIC OCEAN

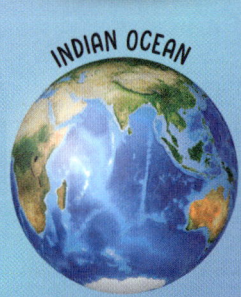
INDIAN OCEAN

SOUTHERN OCEAN

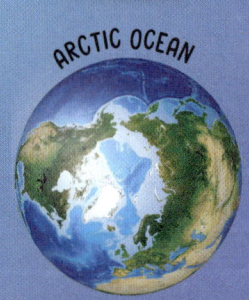
ARCTIC OCEAN

1 Earth's oceans formed about 3.8 BILLION YEARS AGO.

Before then, Earth was too hot for liquid water to exist.

2 THEY CONTAIN about 1.35 BILLION cubic km (324 million cubic miles) of WATER in total.

That's 97 per cent of all the water on Earth.

3 The LARGEST and DEEPEST ocean IS THE PACIFIC OCEAN.

It is so big that all of the world's continents could fit inside it!

4 About 3.5 PER CENT of the WEIGHT of seawater is from SALTS.

Most salts enter the oceans from rivers flowing into them.

5 MOST of the world's ANIMALS LIVE IN THE OCEAN.

Measured by weight, marine creatures make up nearly 80 per cent of animals.

30
June
STATES OF MATTER

1 There are THREE MAIN STATES OF MATTER: SOLIDS, LIQUIDS, and GASES.

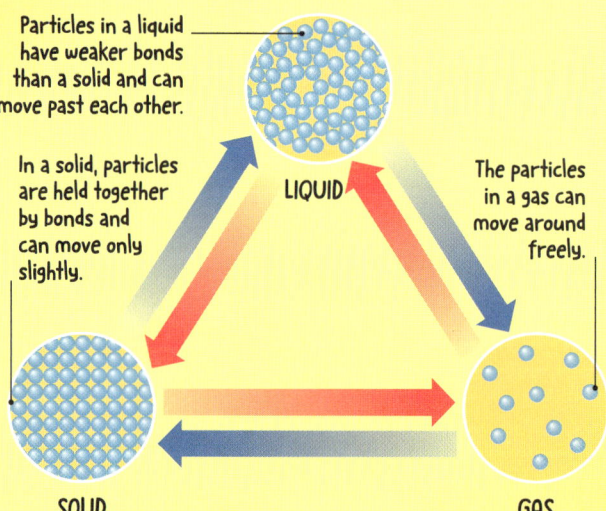

Particles in a liquid have weaker bonds than a solid and can move past each other.

In a solid, particles are held together by bonds and can move only slightly.

LIQUID

The particles in a gas can move around freely.

SOLID

GAS

2 All matter contains TINY PARTICLES (atoms or molecules), which BEHAVE DIFFERENTLY IN DIFFERENT STATES.

3 SUBSTANCES CAN CHANGE from one state to another.

When liquid water freezes, it turns into solid ice.

4 SUBLIMATION is when A SOLID TURNS INTO A GAS.

Usually, it would become a liquid first.

5 HEATING or COOLING a substance causes it to CHANGE STATE.

It happens at different temperatures for different substances - tungsten doesn't melt until it reaches temperatures of 3,414°C (6,177°F)!

1

STEM CELLS are special cells that can TURN INTO MANY DIFFERENT TYPES OF CELL.

RED BLOOD CELLS

SKIN CELLS

NERVE CELLS

Embryonic stem cells can become any type of cell, allowing them to form all the different parts of the developing embryo.

WHITE BLOOD CELLS

EMBRYONIC STEM CELLS

PLACENTAL CELLS

These form the placenta, an organ that feeds an embryo as it grows.

MUSCLE CELLS

FAT CELLS

These are some of the most common type of cells in your body.

EPITHELIAL CELLS

2

The MOST POWERFUL STEM CELLS are those in developing embryos (see p177). These can become any other type of cell.

1
July
STEM CELLS

5

BLOOD CANCERS can be TREATED WITH STEM CELLS, and they could be used to treat more conditions in the future!

New drugs can also be tested on stem cells, and watching them grow into specific types of cells can help scientists understand how diseases are caused.

ADULT STEM CELLS

Adult stem cells are called multipotent as they can turn into several different types of cell, but not all.

Stem cells found in bone marrow make blood cells.

3

Some stem cells are FOUND IN ADULTS.

They occur in places such as the brain, heart, eyes, teeth, and a tissue inside your bones called bone marrow.

WHITE BLOOD CELLS

RED BLOOD CELLS

4

Stem cells in the body usually REPLACE DEAD or DAMAGED CELLS in specific tissues.

135

1

PHOTOSYNTHESIS is a chemical process that PLANTS USE TO MAKE FOOD from water and air, using the energy in sunlight.

Simple organisms called algae can photosynthesize too.

2

It takes place in special structures inside plant cells called CHLOROPLASTS.

These contain a green pigment called chlorophyll that can absorb light energy.

Sunlight provides the energy to power the chemical reactions of photosynthesis.

Carbon dioxide from the air is absorbed through tiny holes in leaves.

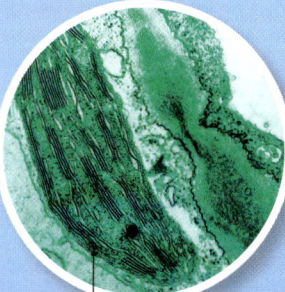

This microscope picture shows a chloroplast magnified hundreds of times.

3

The ENERGY FROM SUNLIGHT is used to combine water and carbon dioxide into GLUCOSE (a sugar) and OXYGEN.

Plants use glucose as an energy store and as a building block to make all the other substances they need to grow.

Waste oxygen gas escapes back into the air.

The roots of a plant draw up water from the soil.

4

Photosynthesis POWERS most of the world's FOOD CHAINS (see p164).

Without plants making energy this way, there would be no energy to pass up the chain.

5

As well as plants and algae, a tiny number of ANIMALS CAN CARRY OUT PHOTOSYNTHESIS!

Some sea slugs can keep chloroplasts inside them after eating algae. And many sea creatures, including corals and anemones, contain algae that photosynthesize.

2

July

PHOTOSYNTHESIS

3
July
FINGERPRINTS

1 FINGERPRINTS are swirly patterns made by TINY RIDGES OF SKIN on our fingertips.

We also have swirly skin patterns on our toes, feet, and palms.

2 FINGERPRINTS ARE UNIQUE. No two people, even twins, have the same ones.

This is why fingerprinting is so useful for identifying people.

3 MONKEYS, APES, and GORILLAS also have unique fingerprints, and so do KOALAS.

These marsupials are the only non-primate to have them.

Koalas' hands have two thumbs each.

You leave greasy fingerprints when you touch anything.

4 Fingerprints were first used to SOLVE CRIMES in the early 1900s.

When people touch things, tiny droplets of grease from their fingertips leave a copy of their unique fingerprints behind.

5 They are NOT the only unique bits of you.

People can also be identified by their iris (the coloured part of the eye), their teeth, their voice, and the shape of their ears.

Round patterns are called whorls.

Bendy patterns are called loops or arches.

---

1 A DAM is a structure that HOLDS BACK FLOWING WATER, forming an artificial pond or lake.

2 The LARGEST LAKE formed by a dam is TWICE THE SIZE OF LUXEMBOURG.

Africa's Lake Kariba took 5 years to fill and covers an area of 5,200 square km (2,000 square miles).

Beavers regularly inspect their dams for leaks, and make repairs if they find any.

3 BEAVERS don't live in dams.

They build dams with sticks to make ponds. Then they build a large stick-nest called a lodge in the pond.

4 The dam at China's JINPING-I POWER STATION is about the SAME HEIGHT AS THE EIFFEL TOWER.

At 305 m (1,000 ft) in height, it is the world's tallest dam.

5 Some modern dams have FISH LADDERS.

These allow migrating fish to swim upstream, over or around the dam, to their breeding sites.

4
July
DAMS

1

OCEAN CURRENTS are constantly MIXING AND PUSHING OCEAN WATER all around the globe.

They flow both at the surface and deep underwater.

2

SURFACE CURRENTS are POWERED BY WINDS blowing over the water.

When wind pushes warm surface water away, cold, nutrient-rich water can rise from the depths.

3

It takes 1,000 YEARS for water to complete a full cycle of the OCEAN CONVEYOR BELT, a global system of ocean currents.

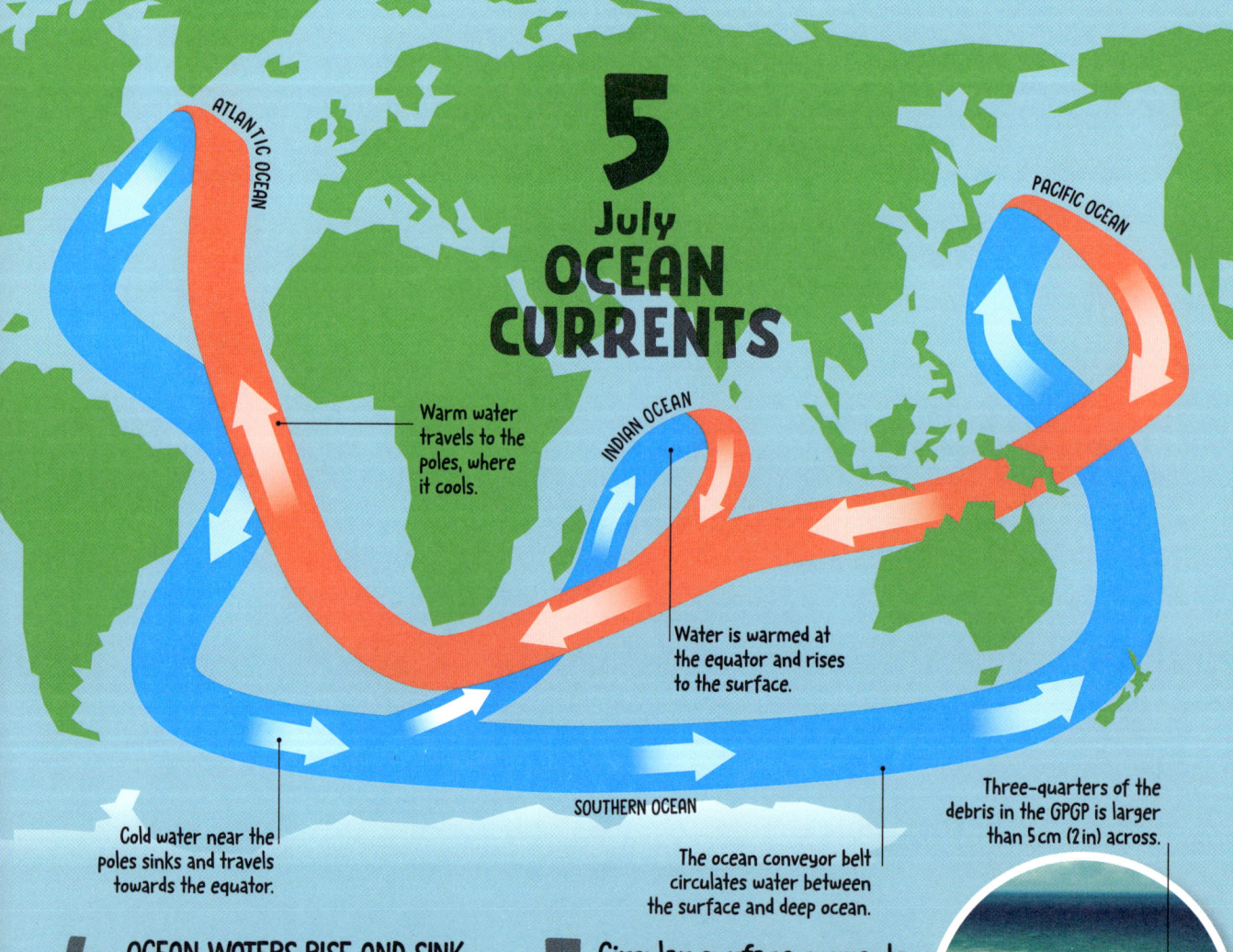

5 July OCEAN CURRENTS

ATLANTIC OCEAN

PACIFIC OCEAN

INDIAN OCEAN

Warm water travels to the poles, where it cools.

Water is warmed at the equator and rises to the surface.

Cold water near the poles sinks and travels towards the equator.

SOUTHERN OCEAN

The ocean conveyor belt circulates water between the surface and deep ocean.

Three-quarters of the debris in the GPGP is larger than 5 cm (2 in) across.

4 OCEAN WATERS RISE AND SINK because of DENSITY (see p169).

Cold, salty water is dense, so it sinks and leaves space at the surface for warm, less salty water to take its place.

5 Circular surface currents created the GREAT PACIFIC GARBAGE PATCH (GPGP), a collection of more than a TRILLION PIECES OF LITTER.

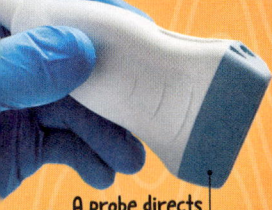

A probe directs ultrasound waves into the body.

1
SOUND WAVES with a frequency TOO HIGH for humans to hear are called ULTRASOUND WAVES.

Any sounds above about 20 kHz are considered ultrasonic.

2
ANIMALS hear different RANGES OF SOUNDS to humans.

Many rodents, bats, frogs, and insects communicate using ultrasonic sounds.

3
ULTRASONIC WAVES can CLEAN medical tools.

The high-frequency vibrations dislodge dirt and germs.

4
ULTRASOUND SCANS are used to SEE INSIDE THE BODY.

Waves sent from a probe bounce off body parts, creating a picture of what's inside.

The probe detects the reflected waves.

5
The scans create a MOVING IMAGE.

They can show how a baby is moving around inside the uterus or how blood flows through the heart.

An image is transmitted to a computer screen.

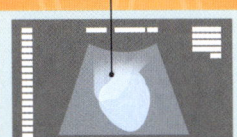

6
July
ULTRASOUND

Stages are jettisoned when their fuel is used up.

7
July
ROCKETS

Eventually, the satellite separates from the final rocket stage.

1
There are about 200 ROCKET LAUNCHES worldwide each year.

The cargo they carry into space is called the payload. Most of them carry satellites.

2
Rockets need SO MUCH FUEL to get off Earth's surface that they are made up of a SERIES OF "STAGES".

3
It only takes about 10 MINUTES for a rocket launching a payload into Earth orbit TO REACH ITS TOP SPEED of about 28,300 km/h (17,600 mph).

4
Rockets are usually LAUNCHED NEAR THE COAST, so that any falling debris will land safely in the sea.

5
The biggest rocket ever built is SpaceX's STARSHIP.

At 121 m (327 ft) tall, it is taller than the Statue of Liberty.

8
July
MIGRATION

Humpback whales make annual migrations from cold polar feeding grounds to breed in more tropical waters.

Atlantic salmon

Monarch butterflies in North America fly south in the autumn, and their descendants return in the spring.

1
Migration is the SEASONAL MOVEMENT of animals from one habitat to another.

Animals migrate to find food, better conditions in winter, or a place to breed.

2
BIRDS, MAMMALS, REPTILES, FISH, AMPHIBIANS, and INSECTS all can migrate.

Their journeys can be across land, sea, and sky.

Large herds of wildebeest migrate across the Serengeti plains each year.

Arctic tern

3
The ARCTIC TERN makes the LONGEST MIGRATION of any animal!

It flies 75,000 km (46,600 miles) all the way from the Arctic to the Antarctic and back every year.

4
ATLANTIC SALMON must SWIM UPRIVER, against the current, to complete their migration.

They start life in rivers, migrate to the sea, and then travel all the way back to where they were born to spawn.

5 Animals instinctively KNOW WHERE TO GO when they migrate.

They can navigate using Earth's magnetic field, smells, or by following the stars and the Sun.

9
July
CUTTING-EDGE MATERIALS

1 Scientists are using ADVANCES in CHEMISTRY and PHYSICS to create new materials.

These can be stronger, lighter, or more sustainable.

2 SELF-HEALING CONCRETE that can FIX ITS OWN CRACKS is being developed!

This could help make buildings last longer.

Photochromatic pigments are used in some sunglasses lenses.

In light, they change colour and darken.

3 Smart materials can change and RESPOND TO LIGHT and heat.

Future homes might use these to automatically adapt and keep us at the right temperature.

4 FUNGI (see p232) and other living things could be made into materials.

Pairs of trainers have been made by growing a mushroom-based alternative to leather.

5 Nanomaterials can be JUST A FEW LAYERS OF ATOMS thick.

Nanomaterials such as titanium dioxide are used in cosmetics or in paints and coatings.

10
July
HOT-AIR BALLOONS

1 Hot-air balloons work by FLOATING.

Hot air is less dense than cold air, so heating the air inside the balloon envelope causes it to rise.

2 The FIRST EVER HOT-AIR BALLOON PASSENGERS were a sheep, a duck, and a rooster.

They were launched in Paris, France, by the Montgolfier brothers in 1783.

3 The AIR INSIDE A BALLOON can reach about 120°C (250°F).

The envelope is made from special heat-resistant material.

4 Hot-air balloons can fly up to around 1,000 m (3,000 ft) ABOVE THE GROUND.

5 To land, the pilot REDUCES OR SWITCHES OFF the burner.

As the air inside the envelope cools, the balloon starts to sink.

1 Earth formed 4.5 BILLION YEARS AGO (BYA) and it took a BILLION YEARS FOR LIFE TO EVOLVE.

Prehistoric animals didn't all exist at the same time – they evolved over billions of years and lived in different periods.

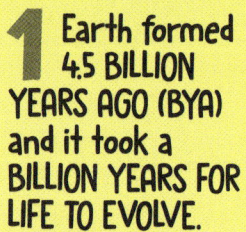

First simple life forms evolve — 3.7 BYA

Evolution of animals occurs — 541 MYA

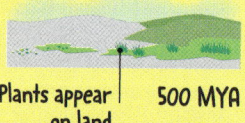

Plants appear on land — 500 MYA

Dinosaurs first evolve

245 MYA

Modern humans (*Homo sapiens*) evolve

300,000 YA

2 The CARBONIFEROUS PERIOD (359–299 MYA) is known for its VAST, SWAMPY FORESTS and OCEANS FULL OF MARINE LIFE.

Lepidodendron plants could reach 30 m (100 ft) tall.

Meganeura had a 75-cm (2.5-ft) wingspan.

3 GIANT INSECTS evolved in the Carboniferous Period. They may have grown so large because the AIR WAS MORE OXYGEN-RICH THAN TODAY.

4 PTEROSAURS were the FIRST ANIMALS AFTER INSECTS TO FLY. These flying reptiles lived alongside dinosaurs.

Quetzalcoatlus had a wingspan of 12 m (40 ft) – the size of a fighter jet.

Short-necked plesiosaurs called pliosaurs could reach 15 m (50 ft) in length.

5 PLESIOSAURS WERE GIANT REPTILES that lived in the oceans during the time of the dinosaurs.

They had four large, powerful flippers.

11
July
PREHISTORIC LIFE

1

A GEYSER is a NATURAL FOUNTAIN that SHOOTS HOT WATER AND STEAM into the air.

Minerals (see p180) underground dissolve into the hot water, making it very mineral-rich.

2

Geysers appear when water in spaces underground is HEATED BY MAGMA (see p14).

Pressure builds up until the water is violently expelled.

3

Water rising up a geyser from deep underground can be THREE TIMES HOTTER THAN THE BOILING POINT OF WATER.

Some geysers erupt every few minutes, while others have intervals of many hours.

12 July GEYSERS

4

Geysers occur in areas with lots of VOLCANIC ACTIVITY or on BOUNDARIES BETWEEN TECTONIC PLATES (see p72).

3. The pressure eventually forces a jet of heated water and steam to erupt into the air through a vent.

Iceland lies on a boundary between two tectonic plates and is home to at least 20 active geysers.

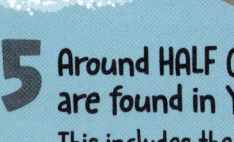

2. As the water heats up and turns to steam, pressure builds up inside the underground chambers.

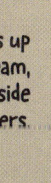

1. Hot magma underground heats up the water.

5

Around HALF OF ALL THE WORLD'S GEYSERS are found in Yellowstone National Park, USA.

This includes the world's tallest geyser, Steamboat, which can shoot water 91 m (300 ft) into the air.

1

Saturn is the FURTHEST PLANET you can see from Earth WITH THE NAKED EYE.

But you can't see its rings. In 1610, when Italian scientist Galileo Galilei studied Saturn through a telescope, he didn't know they were rings and described them as "ears".

2

SATURN'S RINGS are made of BILLIONS of chunks of ICE.

The rings stretch more than 280,000 km (174,000 miles) from Saturn, but are very thin – some are just 10 m (33 ft) deep!

3

YOU COULD FIT 764 EARTHS inside Saturn.

Saturn spins so fast that it bulges at the equator.

The gaps between the rings are made by orbiting moons.

The planet has eight main rings. The ice in the rings reflects the Sun's rays, which is why they are visible with a telescope.

5

Saturn has more moons than any other planet: 146 AND COUNTING...

Its biggest moon, Titan, is bigger than the planet Mercury.

The moons are held in orbit around Saturn by its gravity.

4

If you could build a big enough bath, SATURN WOULD FLOAT.

This is because Saturn is mostly made of the gases hydrogen and helium, with some liquid further in, and a small rocky core. This means that the whole planet has a density (see p169) lower than water, and would float.

14
July
VACUUMS

1 A VACUUM is somewhere with ABSOLUTELY NOTHING IN IT – NOT EVEN AIR.

In the real world, this situation is impossible to create. Instead we have "near vacuums".

2 The CLOSEST thing we know of to a natural VACUUM is SPACE.

Gravity pulls space objects towards each other, leaving huge empty areas in between, which contain only a few atoms.

3 SOUNDS CANNOT TRAVEL THROUGH A VACUUM.

This is because sounds travel when particles vibrate. In a vacuum there are no particles, so the sound can't go anywhere!

4 VACUUM CLEANERS use a partial VACUUM TO CLEAN.

The partial vacuum is created by a fan, which pulls air and dirt into the machine.

5 DRINKS FLASKS use VACUUMS to keep WARM DRINKS WARM and COOL DRINKS COOL.

The warm or cool liquid goes inside, with a vacuum chamber wrapped around it.

The flask is like two bottles, one inside the other.

The vacuum in between stops most heat from getting in or out.

15
July
CRYSTALS

1 Crystals are solid materials that have their ATOMS (see p53) ARRANGED in REGULAR, REPEATING 3D PATTERNS.

2 CRYSTALS FORM IN DIFFERENT WAYS.

Some form when liquid rock cools down. Others, such as salt crystals, form from mineral-rich solutions of water.

3 Usually, THE MORE SLOWLY A CRYSTAL FORMS, THE BIGGER IT CAN GET.

4 Most GEMSTONES (see p235) are CUT from MINERALS with CRYSTAL STRUCTURES.

They are shaped to make them sparkle, then set into pieces of jewellery.

5 ICE CRYSTALS FORM WHEN WATER FREEZES.

When water vapour in the air freezes, it forms snowflakes. If the vapour is on a cold surface, it freezes to form frost.

7. Finally, the SM is released and the CM splashes into the ocean.

1. The *Saturn V* rocket launches Apollo into Earth orbit.

2. The first part of the rocket falls away.

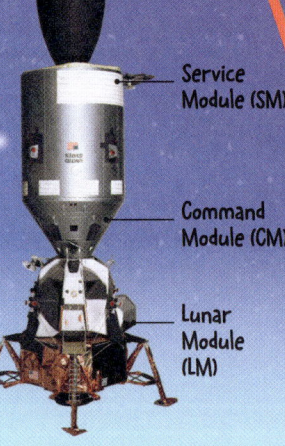

Service Module (SM)

Command Module (CM)

Lunar Module (LM)

1 The Soviet Union's uncrewed *Luna 1* was the FIRST SPACECRAFT TO REACH THE MOON, IN 1959.

2 There have been MORE THAN 140 MISSIONS to the Moon so far.
Only nine of those missions sent people there. All crewed missions were part of NASA's Apollo programme.

3. The SM, CM, and LM separate from the rocket.

3 It took the Apollo missions around THREE DAYS TO FLY FROM EARTH TO THE MOON.
NASA, the USA's space agency, used the *Saturn V* rocket to send their Apollo spacecraft and three astronauts to the Moon.

4. The LM detaches and carries the astronauts to and from the Moon's surface.

4 All THE APOLLO SPACECRAFT were made up of THREE PARTS.
Crew lived in the Command Module; the Service Module held fuel and power; the Lunar Module landed on the Moon.

6. The SM and CM head back towards Earth.

16
July
JOURNEY TO THE MOON

5 It took just 8 HOURS AND 35 MINUTES for NASA's *New Horizons* spacecraft to REACH THE MOON in 2006, on its way to Pluto.

5. The LM returns the astronauts to the CM and is then abandoned in space.

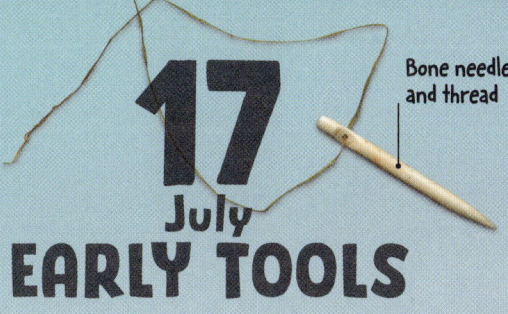

17 July
EARLY TOOLS

Bone needle and thread

1

Humans have been making STONE TOOLS for at least 2.6 MILLION YEARS.

2

The FIRST TOOLS humans made were HAND AXES.

They were made from a type of stone called flint.

This flint arrowhead would have been tied to a stick to make an arrow for hunting animals.

Goose feather brush

Harpoon

Flint hand axe

3

Later, people added SHARPENED PIECES OF FLINT to STICKS to create SPEARS and ARROWS.

4

People then used animal IVORY, BONE, and HORN to make a range of other tools, like NEEDLES and HARPOONS.

5

In around 3,000 BCE, people learned to make BRONZE – by adding tin to copper – which was A STRONG METAL FOR TOOLS AND WEAPONS.

18 July
MARSUPIALS

1

Marsupials are MAMMALS (see p98). Most marsupials have POUCHES.

Most marsupials live in Australia. They include kangaroos and koalas.

2

MARSUPIAL BABIES ARE CALLED JOEYS.

They are born with very underdeveloped bodies, so they crawl up into their mother's pouch to stay and drink milk until they are ready for the world.

3

GLIDERS are marsupials that can TAKE TO THE SKY.

They use big flaps of skin on their sides to help them glide from tree to tree.

4

When a VIRGINIA OPOSSUM'S babies grow too big for its pouch, they cling to their MOTHER'S BACK, so she can still carry them around.

5

A kangaroo's POWERFUL LEGS and LARGE FEET help it hop up to 9 m (30 ft) to ESCAPE danger.

As a joey gets older, it gets out and back into the pouch. Eventually, it will get too big to fit!

19
July
LEAVES

1 Leaves are like a PLANT'S SOLAR PANELS.

They use energy from sunlight to make food in a process called photosynthesis (see p136).

2 TINY HOLES, called STOMATA, underneath a leaf allow the GASES needed for photosynthesis to PASS IN AND OUT.

Most photosynthesis happens in tall, narrow cells in the top layer of a leaf. They are packed with chlorophyll.

Guard cells either side of the stomata swell to open the stomata during the day so photosynthesis can occur.

Bundles of veins transport water, sugar, and minerals throughout a plant.

Stomata

3 Leaves are GREEN because they contain CHLOROPHYLL.

This pigment is essential for trapping energy from the Sun for photosynthesis.

4 Leaves CHANGE COLOUR when LIGHT LEVELS DIP in autumn.

The lack of light causes the green chlorophyll pigment to break down, leaving other colourful pigments behind, such as yellow, orange, and red.

A protective surface called the waxy cuticle prevents leaves from losing excess water.

Poinsettia plants have bright-red bracts that look like large petals.

5 Some plants have developed VERY UNUSUAL LEAVES.

Cacti leaves evolved into spines to prevent water loss in the hot desert. Other plants have petal-like leaves called bracts that attract pollinators.

Apollo 11 astronauts (from left to right): Edwin "Buzz" Aldrin, Michael Collins, and Neil Armstrong

20 July MOON LANDINGS

5 There are about 180 TONNES (400,000 lb) of JUNK on the Moon.

That includes 96 bags of human waste, four rovers, several flags, two golf balls, and a family photo.

1 There have been SIX MOON LANDINGS and 12 PEOPLE have walked on the Moon... so far.

The first landing was the US Apollo 11 mission, on 20 July 1969.

4 Astronauts have brought back more than 2,000 samples of MOON ROCK, PEBBLES, SAND, and DUST.

Scientists still use the samples to find out more about the structure and formation of the Moon.

2 WALKING on the Moon is tricky!

This is because its pull of gravity (see p75) is a sixth of the force you experience on Earth, so it's easier to hop and bounce than to walk.

The Apollo 11 Lunar Module was the first ever vehicle with a crew to land on the Moon.

An astronaut can jump six times higher on the Moon than on Earth.

3 Astronauts used LUNAR ROVERS to explore the Moon's SURFACE.

They had solid wheels to cope with the rocky terrain and could reach speeds of 18 km/h (11 mph).

1 Doctors use incredible MACHINES TO SEE INSIDE YOUR BODY and diagnose illnesses.

MRI scanners use powerful magnets and radio waves to create detailed images of the body, including the brain, bones, and internal organs.

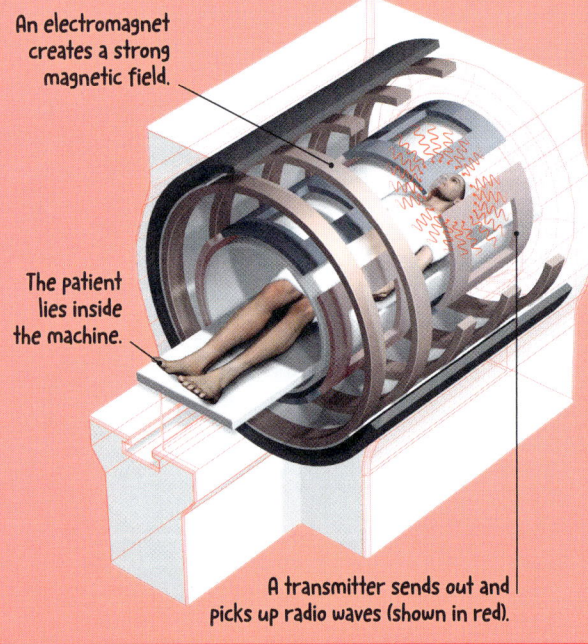

An electromagnet creates a strong magnetic field.

The patient lies inside the machine.

A transmitter sends out and picks up radio waves (shown in red).

2 SCIENTISTS CAN NOW "EDIT" (CHANGE) GENES that cause some diseases.

Faulty genes (see p29) are removed, repaired, or replaced with healthy ones.

Strand of DNA

1. Gene-cutting enzyme matches the target section of DNA.

2. Enzyme finds matching sequence.

3. Faulty gene is located and removed.

4. Healthy gene is inserted.

21
July
MODERN MEDICINE

3 Biotechnologists have developed DEVICES THAT CAN RESTORE SENSES.

Cochlear implants help some deaf people to hear. Advanced artificial limbs (see p63) are being developed that simulate the sensation of touch.

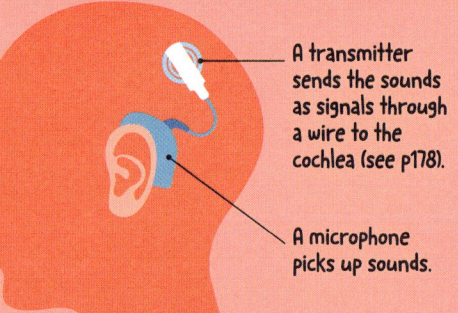

A transmitter sends the sounds as signals through a wire to the cochlea (see p178).

A microphone picks up sounds.

5 Unspecialized cells called STEM CELLS (see p135) have HUGE POTENTIAL IN MEDICINE.

Stem cell therapy uses stem cells to grow healthy tissue and treat illnesses.

Human stem cells under a microscope

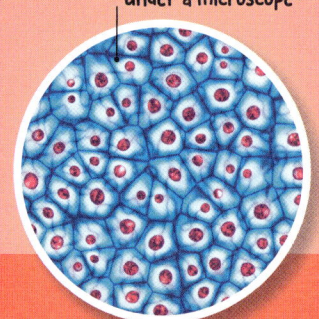

4 Advances in BIOENGINEERING mean that SKIN CAN BE GROWN IN A LABORATORY for treating burn victims.

Research is also under way into the use of 3D bioprinting to print skin directly onto a wound, saving time and reducing the risk of infection.

22
July
SEASHELLS

1 Shells are hard structures that provide animals with ARMOUR-LIKE PROTECTION and a sturdy surface for them to ATTACH their MUSCLES.

2 Some seashells have GROWTH LINES a little like tree rings that can tell you HOW OLD THE ANIMAL IS.

3 The shiny inside layer of shells, called NACRE or MOTHER-OF-PEARL, helps make a shell CRACK-PROOF.

4 Most coiled shells are "DEXTRAL", which means they OPEN TO THE RIGHT.

Left-opening "sinistral" shells are very rare.

5 Seashells are good at AMPLIFYING SOUND.

They have been used as musical instruments across the globe and some cultures have even used shells as currency.

23
July
ICE SKATING

1 The first ice skates were probably MADE OF ANIMAL BONES.

The people of Scandinavia used them to cross frozen lakes in winter.

The blades of modern ice skates are made of steel.

2 THE FASTER YOU SKATE, THE SLIPPIER ICE GETS.

Skates create a force called friction, which melts the ice.

3 Skating on ice that's too cold IS LIKE SKATING ON CONCRETE.

When ice is colder than –30°C (–22°F), skates can't melt it and it stays rock solid.

4 SKATING ON LAKES is POSSIBLE only because WATER DOES SOMETHING ODD.

When water freezes, it gets lighter and floats. That's why lakes freeze on top in winter.

5 The fastest a skater has ever spun is a dizzying SIX REVOLUTIONS A SECOND!

Skaters can spin faster by tucking their arms in or holding them up, to reduce their radius.

24
July
ELEMENTS

1 An ELEMENT is a PURE SUBSTANCE that is made from only ONE TYPE OF ATOM (see p53).

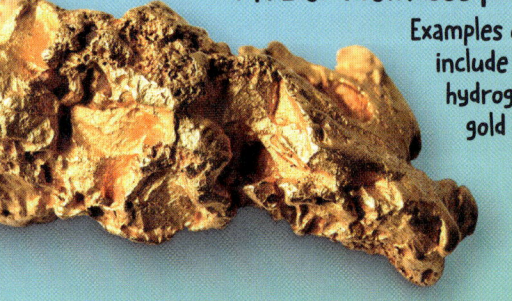

Examples of elements include oxygen, hydrogen, carbon, gold (left), and lead.

2 Most elements DO NOT OCCUR NATURALLY in their PURE FORM.

To be isolated, they have to be extracted from minerals.

3 There are 118 DIFFERENT ELEMENTS, which are listed in the periodic table.

Oganesson – the most recent element to be discovered – was added to the table in 2015.

4 Only TWO ELEMENTS exist naturally on Earth as LIQUIDS – MERCURY and BROMINE.

Liquid mercury is very slippery.

5 German chemist Hennig Brand DISCOVERED the element PHOSPHORUS in 1669 BY BOILING HIS OWN URINE!

1 Russian chemist Dmitri MENDELEEV INVENTED THE PERIODIC TABLE in 1869.

It was a way of organizing all the known elements.

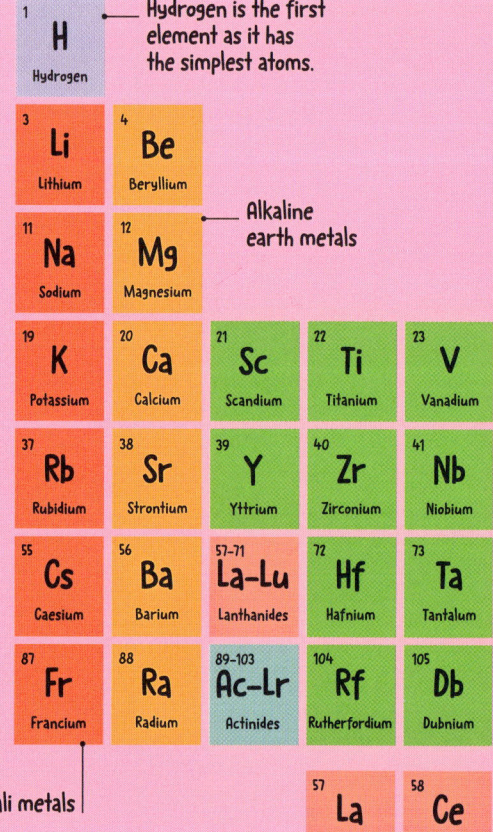

1 H Hydrogen		

Hydrogen is the first element as it has the simplest atoms.

3 Li Lithium	**4** Be Beryllium
11 Na Sodium	**12** Mg Magnesium

Alkaline earth metals

19 K Potassium	**20** Ca Calcium	**21** Sc Scandium	**22** Ti Titanium	**23** V Vanadium
37 Rb Rubidium	**38** Sr Strontium	**39** Y Yttrium	**40** Zr Zirconium	**41** Nb Niobium
55 Cs Caesium	**56** Ba Barium	**57-71** La-Lu Lanthanides	**72** Hf Hafnium	**73** Ta Tantalum
87 Fr Francium	**88** Ra Radium	**89-103** Ac-Lr Actinides	**104** Rf Rutherfordium	**105** Db Dubnium

Alkali metals

57 La Lanthanum	**58** Ce Cerium
89 Ac Actinium	**90** Th Thorium

The lanthanides are sometimes known as rare earth metals.

The actinides contain many recently discovered elements, such as Einsteinium, that are named after famous scientists.

3 GROUPS of elements have SIMILAR PROPERTIES.

These are shown in different colours on the table.

PERIODIC TABLE

2

ELEMENTS ARE ORDERED from 1 to 118 based on their ATOMIC NUMBER – the number of protons in each of their atoms.

This column is the noble gases, which all exist as a gas at room temperature.

Carbon group

Nitrogen group

Oxygen group

Halogens

Transition metals are good conductors of heat and electricity.

The Boron group contains metals and metalloids.

													2 He Helium
							5 B Boron	6 C Carbon	7 N Nitrogen	8 O Oxygen	9 F Fluorine	10 Ne Neon
							13 Al Aluminium	14 Si Silicon	15 P Phosphorus	16 S Sulfur	17 Cl Chlorine	18 Ar Argon
24 Cr Chromium	25 Mn Manganese	26 Fe Iron	27 Co Cobalt	28 Ni Nickel	29 Cu Copper	30 Zn Zinc	31 Ga Gallium	32 Ge Germanium	33 As Arsenic	34 Se Selenium	35 Br Bromine	36 Kr Krypton
42 Mo Molybdenum	43 Tc Technetium	44 Ru Ruthenium	45 Rh Rhodium	46 Pd Palladium	47 Ag Silver	48 Cd Cadmium	49 In Indium	50 Sn Tin	51 Sb Antimony	52 Te Tellurium	53 I Iodine	54 Xe Xenon
74 W Tungsten	75 Re Rhenium	76 Os Osmium	77 Ir Iridium	78 Pt Platinum	79 Au Gold	80 Hg Mercury	81 Tl Thallium	82 Pb Lead	83 Bi Bismuth	84 Po Polonium	85 At Astatine	86 Rn Radon
106 Sg Seaborgium	107 Bh Bohrium	108 Hs Hassium	109 Mt Meitnerium	110 Ds Darmstadtium	111 Rg Roentgenium	112 Cn Copernicum	113 Nh Nihonium	114 Fl Flerovium	115 Mc Moscovium	116 Lv Livermorium	117 Ts Tennessine	118 Og Oganesson

59 Pr Praseodymium	60 Nd Neodymium	61 Pm Promethium	62 Sm Samarium	63 Eu Europium	64 Gd Gadolinium	65 Tb Terbium	66 Dy Dysprosium	67 Ho Holmium	68 Er Erbium	69 Tm Thulium	70 Yb Ytterbium	71 Lu Lutetium
91 Pa Protactinium	92 U Uranium	93 Np Neptunium	94 Pu Plutonium	95 Am Americium	96 Cm Curium	97 Bk Berkelium	98 Cf Californium	99 Es Einsteinium	100 Fm Fermium	101 Md Mendelevium	102 No Nobelium	103 Lr Lawrencium

4 The Periodic Table WASN'T COMPLETE WHEN IT WAS MADE.

Only 63 elements were on it, but Mendeleev left space for more that might be discovered.

5 NOT ALL of the elements on the table EXIST on Earth NATURALLY.

Some are very unstable and have been created only temporarily in labs.

26
July
WATER CYCLE

1 WATER on Earth is ALWAYS ON THE MOVE – it travels around the planet in a CONTINUOUS CYCLE.

2 The SUN POWERS the water cycle.
It warms the water in seas, rivers, and lakes, which evaporates (becomes water vapour) and rises into the sky. Plants also release water vapour.

3 As it COOLS in the sky, WATER VAPOUR BECOMES LIQUID AGAIN.

This is called condensation. The liquid water droplets build up to form clouds (see p46).

Clouds release water as rain when the water droplets become too heavy.

Some water is stored as ice in glaciers.

CONDENSATION

EVAPORATION

PRECIPITATION

Water flows downhill until it meets a river.

4 Eventually, the water FALLS BACK TO EARTH as PRECIPITATION, such as rain, snow, or hail.

This water flows into rivers, lakes, and oceans, and soaks into the ground, and the cycle begins again.

Water that seeps underground is called groundwater. It flows through tiny spaces between the rocks and soil.

5 It can take anywhere from a FEW HOURS to THOUSANDS OF YEARS for a DROP OF WATER to complete one cycle.

27 July
CARNIVOROUS PLANTS

1
There are MORE THAN 600 SPECIES of carnivorous plant, which ATTRACT, TRAP, and DIGEST animals.

2
The plants usually live in POOR-QUALITY SOIL, so they get their NUTRIENTS from LIVING CREATURES.

3
Some PITCHER PLANTS are big enough to trap SMALL ANIMALS such as rats, frogs, and even birds.

4
SUNDEWS attract prey with SWEET, STICKY DROPLETS on their leaves.

Once prey is stuck, the plant curls up to digest its meal.

5 LOW'S PITCHER PLANT feeds on SHREW DROPPINGS!

As a shrew feeds on the plant's lid, it poos into the pitcher, delivering nutrients to the plant.

Snap traps, such as the Venus flytrap, have sensitive trigger hairs inside the trap.

When hairs are triggered by prey, the leaf blades snap shut in one-tenth of a second.

28 July
STAR BIRTH

1
Stars are BORN in MOLECULAR CLOUDS: huge, cold clouds of GAS AND DUST.

2
It can take MILLIONS OF YEARS for new stars to FORM.

3
Stars are born when THE GAS BEGINS TO CLUMP.

Gravity (see p75) causes these clumps to collapse and shrink.

4
Eventually, the clumps become SO HOT AND DENSE that they IGNITE TO FORM STARS.

New bright, young stars light up the gases of the Orion Nebula.

5
The ORION NEBULA is a star birth region that can be seen using BINOCULARS.

The nebula is 1,500 light years away and lies in the Orion constellation.

1 The Moon DOESN'T PRODUCE ITS OWN LIGHT – we see it because it REFLECTS THE SUN'S LIGHT.

2 We don't always see the Moon as A FULL CIRCLE in the sky.

The Sun always lights up one side of the Moon, but because the Moon is orbiting Earth, we see different amounts of its lit side throughout the month.

MOON

THIRD QUARTER

From Earth, we can see only part of the Moon's lit-up face.

WANING CRESCENT

Sunlight falls on the side of the Moon that faces the Sun.

The Moon's orbit around Earth

WANING GIBBOUS

The Moon is described as "waning" when its visible area decreases.

EARTH

29
July
LUNAR CYCLES

SUNLIGHT

NEW MOON

FULL MOON

WAXING CRESCENT

The Moon is described as "waxing" when its visible area increases.

FIRST QUARTER

Only half of the Moon's lit portion is visible from Earth.

WAXING GIBBOUS

3 The DIFFERENT SHAPES of the Moon are called LUNAR PHASES.

A full cycle of lunar phases lasts 29.5 days.

4 We see a FULL MOON when it is on the opposite side of Earth.

This is because sunlight illuminates the whole of the Moon's near side.

5 A NEW MOON occurs when the Moon is positioned BETWEEN EARTH AND THE SUN.

This is because the sunlight cannot fall on the side of the Moon that we see from Earth.

30 July
PLASTICS

1 Plastics are SYNTHETIC (HUMAN-MADE) MATERIALS that are POLYMERS (see p127).

2 There are DIFFERENT TYPES OF PLASTICS.
They range from polyethylene, which is used for plastic bags, to hard PVC used in pipes and building materials.

3 Plastics became POPULAR in the 20th CENTURY because they were lightweight, strong, and cheap to make.

4 It can take UP TO 500 YEARS for some plastics to DECOMPOSE (break down).
Each year, we produce around 400 million tonnes of plastic waste.

5 There could be PLASTIC INSIDE YOUR BODY!
Plastics break down into microplastics less than 5 mm (0.2 in) long that can get into food, water, and ourselves.

31 July
THE INTERNET

1 The internet is a GLOBAL NETWORK OF COMPUTERS that share information.

2 The WORLD WIDE WEB is NOT THE SAME as the internet.
The internet is what connects devices together. The World Wide Web is all the websites you can view on the internet.

3 The FIRST WEBPAGE was launched in 1991.
It was created by British scientist Tim Berners-Lee, who invented the World Wide Web.

4 5.45 BILLION PEOPLE use the internet.
That is around 66 per cent of the world's population!

5 Internet signals TRAVEL UP INTO SPACE and even deep UNDER THE SEA!
Satellites orbiting Earth and undersea cables buried under the seabed transmit internet data.

1

The CLIMATE of a place IS ITS TYPICAL WEATHER, averaged over many years.

Climate zones are major regions of Earth with a distinctive climate.

2

The CLOSER YOU GET to EARTH'S EQUATOR, the WARMER THE CLIMATE is.

And the closer you are to the poles, the colder the climate is.

3

MANY FACTORS AFFECT THE CLIMATE A PLACE HAS.

These include mountains, valleys, how near the sea is, how warm the sea is, and which way the wind usually blows.

1 August CLIMATE ZONES

POLAR ZONE

The polar zone is cold all year.

TEMPERATE ZONE

SUBTROPICAL ZONE

TROPIC OF CANCER

The temperate zone has four seasons: spring, summer, autumn, and winter.

TROPICAL ZONE

EQUATOR

TROPIC OF CAPRICORN

SUBTROPICAL ZONE

The subtropical zone has hot summers but mild winters.

The tropical zone is warm all year, but some parts have wet and dry seasons.

TEMPERATE ZONE

POLAR ZONE

4

A MICROCLIMATE is an unusual CLIMATE in A VERY SMALL AREA.

Gardeners use greenhouses to create a warm microclimate so they can grow plants that wouldn't thrive outside.

5

CLIMATE CHANGE is the INCREASE in OUR PLANET'S average TEMPERATURE.

It's caused by the pollution of Earth's atmosphere with carbon dioxide gas and other gases released by human activity.

ICEBERGS

1 ICEBERGS FORM WHERE ICE SHEETS OR GLACIERS MEET THE OCEAN.

Deep cracks (crevasses) form due to the ice movement and the weight of the shelf.

SUN'S HEAT

Ice breaking away from an ice sheet or glacier is called calving.

2 Ocean currents and the weight of the ice cause the ICE TO CRACK, and PIECES TO BREAK OFF and FLOAT AWAY AS ICEBERGS.

About 90 per cent of an iceberg lies hidden beneath the waves.

ICE FLOW

ICE SHELF

MELTING

Warm water from the ocean depths flows in under the ice shelf and melts the ice.

WARM WATER

Stripes can be blue, green, yellow, brown, or black.

Blue stripes form when water gets trapped in gaps in the ice and freezes too fast for bubbles to form.

3 SOME ICEBERGS ARE STRIPEY.

4 The BIGGEST-EVER ICEBERG was LARGER than BELGIUM!

It was estimated to be 335 km (208 miles) long and 97 km (60 miles) wide.

5 Icebergs LESS THAN 5 m (16 ft) WIDE that BREAK OFF LARGER ICEBERGS are called GROWLERS.

3
August
THREATENED SPECIES

1 MORE THAN 45,300 species of ANIMALS and PLANTS are considered THREATENED.

The International Union for Conservation of Nature (IUCN) keeps track of these. The IUCN Red List has different categories for how much danger species are in.

LEAST CONCERN
Includes:
- Edelweiss flower
- Cockatiel
- Arabian toad

NEAR THREATENED
Includes:
- Jaguar
- American bison
- Emperor penguin

2 HUMAN ACTIVITY has put many ANIMAL AND PLANT SPECIES AT RISK.

Habitat loss is the main cause. Nearly half the world's habitable land is now used for agriculture, leaving less space for wildlife.

VULNERABLE
Includes:
- Diamondback terrapin
- Polar bear
- Hippopotamus

3 A FEW SPECIES thought to be extinct have BEEN FOUND ALIVE AGAIN.

The terror skink is a lizard that lives only on certain Pacific islands. It was thought to have gone extinct but was rediscovered in 1993.

ENDANGERED
Includes:
- Chimpanzee
- Blue whale
- African wild dog

Once considered a delicacy, this turtle was nearly hunted to extinction.

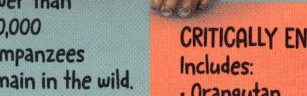
Fewer than 300,000 chimpanzees remain in the wild.

CRITICALLY ENDANGERED
Includes:
- Orangutan
- Black rhinoceros
- European eel

4 About 40 PER CENT of all named PLANT SPECIES are AT RISK OF EXTINCTION.

Scientists reckon that 75 per cent of the species yet to be named or classified are at risk too!

EXTINCT IN THE WILD
Includes:
- Guam kingfisher
- Socorro dove
- Hawaiian crow

Deforestation has put orangutans at risk.

5 Some animals have been brought BACK FROM the BRINK OF EXTINCTION.

Thanks to protection, the population of one-horned rhinos rose from around 200 in the late 20th century to more than 4,000 today.

EXTINCT
Includes:
- Dodo
- Golden toad
- Pinta giant tortoise

The dodo was a flightless pigeon that went extinct in the 17th century.

4 August
GEARS

1 GEARS ARE WHEELS WITH TOOTHED EDGES THAT INTERLOCK TO TURN EACH OTHER.

2 We use gears to CHANGE THE FORCE OR SPEED of rotating parts in a machine.

3 When a SMALL gear DRIVES a LARGE gear, the driven gear TURNS more SLOWLY BUT WITH MORE FORCE.

Input gear

Driven gear

Input gear

Driven gear

4 When a LARGE gear DRIVES a SMALL gear, the driven gear TURNS more QUICKLY BUT WITH LESS FORCE.

5 BIKE GEARS ARE JOINED WITH A CHAIN.

They can magnify the force from your legs (to help you climb a hill) or the speed of your legs (to help you win a race).

5 August
SKYDIVING

1 A SKYDIVER'S SPEED as they hurtle to Earth depends on the BALANCE OF TWO FORCES: GRAVITY AND AIR RESISTANCE.

2 AT FIRST GRAVITY IS GREATER, so the skydiver accelerates downwards at a breathtaking rate.

Terminal velocity in freefall is about 193 km/h (120 mph).

3 After 10 seconds, AIR RESISTANCE is so great that it EQUALS GRAVITY.

The skydiver now falls at a constant speed called terminal velocity.

4 When the PARACHUTE OPENS, AIR RESISTANCE is GREATER than gravity.

The skydiver slows down.

5 AIR RESISTANCE REDUCES as the skydiver slows down UNTIL IT BALANCES GRAVITY AGAIN.

The new, slower terminal velocity is safe for landing.

Terminal velocity with a parachute is about 24 km/h (15 mph).

■ AIR RESISTANCE
■ GRAVITY

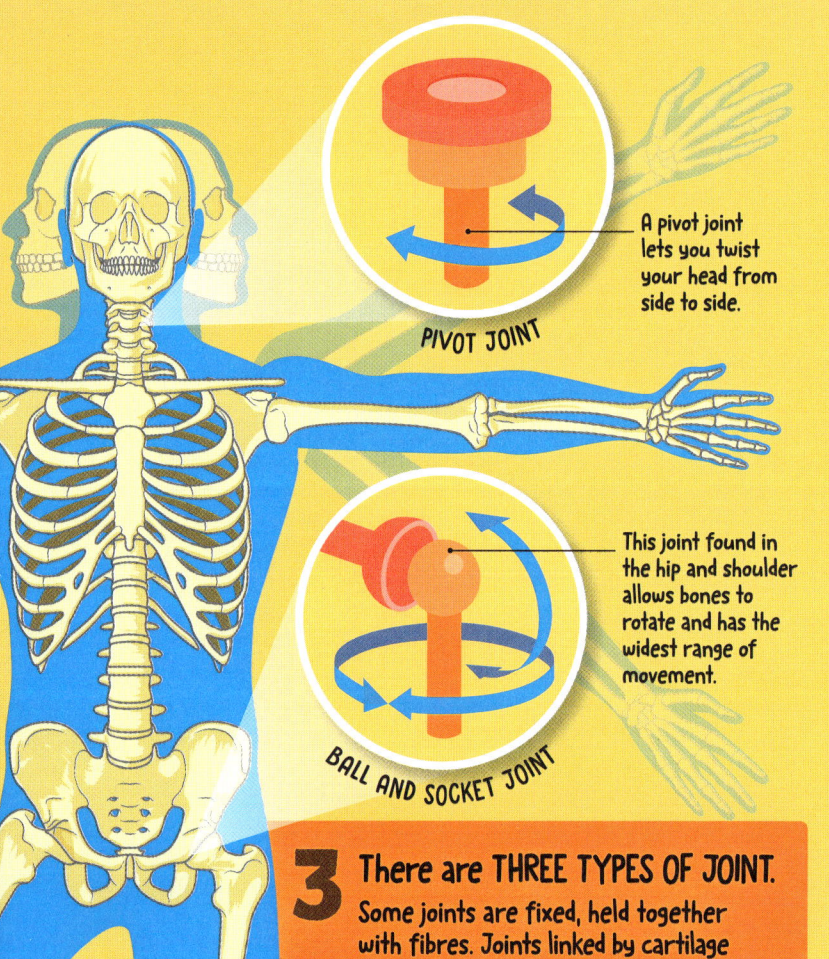

A pivot joint lets you twist your head from side to side.

PIVOT JOINT

This joint found in the hip and shoulder allows bones to rotate and has the widest range of movement.

BALL AND SOCKET JOINT

Hinge joints in the knees and elbows let you move bones in one plane.

HINGE JOINT

6
August
JOINTS

1 A JOINT is where TWO OR MORE BONES MEET.

Joints allow your skeleton and body to move.

2 Each joint allows only a LIMITED RANGE OF MOVEMENT.

When many joints act together, they allow the skeleton to move in complex ways. There are six types of movement, including pivot, ball and socket, and hinge.

3 There are THREE TYPES OF JOINT.

Some joints are fixed, held together with fibres. Joints linked by cartilage allow a little movement, while synovial joints move more freely.

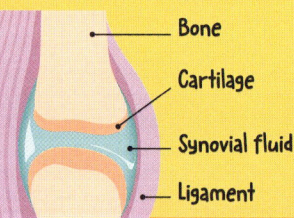

Bone

Cartilage

Synovial fluid

Ligament

4

SYNOVIAL JOINTS are the MOST COMMON joint in the body.

Connective tissues called ligaments hold the bones together, cartilage cushions the ends of the bones, and slippery synovial fluid helps the bones to move smoothly against each other.

5 The HYOID BONE in your throat is the only bone in the body that DOESN'T FORM A JOINT WITH ANOTHER BONE.

7
August
MOUNTAIN LIFE

1

MOUNTAIN HABITATS are found at HEIGHTS MORE THAN 600 m (2,000 ft).

Mount Everest is the highest habitat on Earth, rising to 8,848 m (29,029 ft).

2

The HIGHER THE HABITAT, the COLDER AND WINDIER it is and the LOWER THE AIR PRESSURE.

Low air pressure means there is less oxygen to breathe.

The bar-headed goose takes hundreds of breaths a minute to get the oxygen it needs.

The Himalayan yak has a double-layer coat, helping it withstand temperatures of –40°C (–40°F).

3

The WILD YAK, like many Himalayan animals, has EXTRA-LARGE LUNGS to take in AS MUCH OXYGEN AS POSSIBLE.

Thick fur, snowshoe-like feet, and a camouflaged coat help the snow leopard survive in the mountains.

4

The SNOW LEOPARD is PERFECTLY ADAPTED to its mountain habitat.

However, it is still highly endangered due to habitat loss and human hunters.

5

The HIMALAYAN MARMOT sleeps for up to EIGHT MONTHS A YEAR.

It hibernates inside deep burrows.

1
The SOURCE OF ALL THE ENERGY in a food chain is THE SUN.

Plants and algae use photosynthesis (see p136) to turn sunlight energy into sugars.

8
August
FOOD CHAINS

4. Apex predators hunt smaller animals.

APEX PREDATOR

2
The energy PASSES ALONG THE FOOD CHAIN.

Energy in plants and algae is consumed by animals.

3. Small animals gain energy from eating the herbivores.

SECONDARY CONSUMERS

3
APEX PREDATORS are right at the TOP OF THEIR FOOD CHAIN.

They eat other animals but don't have natural predators themselves.

2. Herbivores take energy from plants.

4
The FURTHER YOU GO up the chain, the FEWER ANIMALS THERE ARE.

Energy is used up along the way, so it takes many plants to support just one apex predator.

1. Plants harness the energy in sunlight to power themselves.

PRIMARY CONSUMERS

5 EVERY SPECIES is part of SEVERAL FOOD CHAINS.

Food chains in an ecosystem link together to form a complex food web.

PRODUCERS

1

Iron filings showing the shape of a magnetic field

The AREA AROUND A MAGNET where its magnetism can affect other materials is known as its MAGNETIC FIELD.

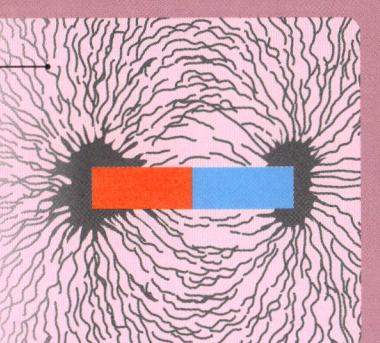

2

MAGNETS ATTRACT and REPEL each other.

Magnets have two poles. If you put the matching poles together, they will repel (push away from) each other. Opposite poles will attract (stick together).

Magnetic field lines

Matching poles repel each other.

Opposite poles attract each other.

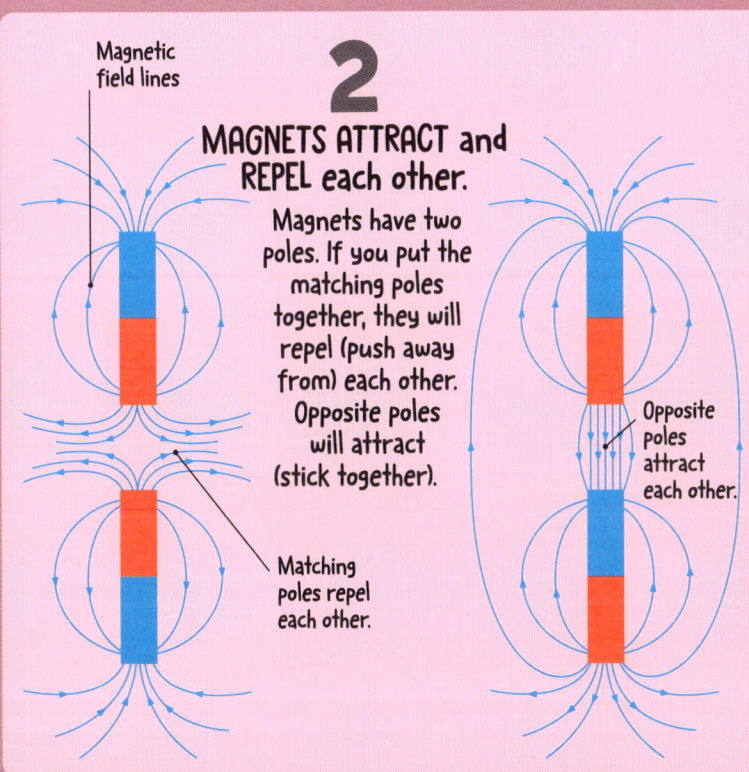

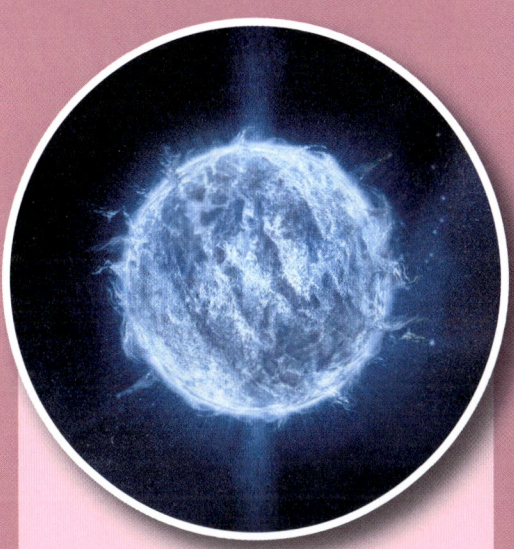

4

NEUTRON STARS have a magnetic field a QUADRILLION TIMES more powerful than Earth's.

3

EARTH is one GIANT MAGNET. Our planet has a huge magnetic field around it.

Earth has two magnetic poles, like a huge magnet.

Earth's magnetic field reaches 64,000 km (40,000 miles) towards the Sun.

5

Iron, nickel, and cobalt are all metals that attract and stick to magnets. However, MOST METALS ARE NOT MAGNETIC.

RECYCLING

1 RECYCLING is a process that TRANSFORMS WASTE PRODUCTS BACK INTO RAW MATERIALS so they can be made into NEW PRODUCTS.

2 HOUSEHOLD RECYCLING is collected and sent to HUGE SORTING CENTRES, where machines sort it into different categories.

Aluminium cans are collected from recycling bins.

The cans are crushed and compressed into large blocks.

Once empty, aluminium cans can be recycled.

Each block is melted down.

The aluminium sheets can be shaped into new products.

The aluminium is rolled out into thin sheets.

3 GLASS, PAPER, PLASTIC, METALS, and CARDBOARD, can all be recycled.

Food waste can be recycled into fertilizer, or burned to provide renewable energy.

5 Almost 70 PER CENT OF ALUMINIUM CANS around the world are RECYCLED.

Aluminium can be recycled many times – almost infinitely.

4 RECYCLING ONE GLASS BOTTLE saves enough energy to POWER A LAPTOP FOR HALF AN HOUR.

166

11
August
OCEAN SLIME

1 Many SEA CREATURES MAKE SLIME and use it in different ways.

Giant larvaceans surround themselves with mucus bubbles that filter food from the ocean.

2 When THREATENED, eel-like HAGFISH release mucus that EXPANDS TO 10,000 TIMES ITS SIZE.

The goo blocks up the throats and gills of predators.

The parrotfish's slime sac may mask its body odour so it can't be found by hungry eels.

3 Ocean molluscs called SEA HARES can secrete STICKY CLOUDS OF TOXIC INK.

It can mask a predator's senses by sticking to their antennae.

4 LIMPETS leave TRAILS OF MUCUS when they move around to feed.

The slime helps them find their way back to their favourite spot.

5 PARROTFISH secrete SLEEPING BAGS OF SLIME.

Scientists think they protect them from parasites and predators while they sleep.

12
August
LIVING ON THE MOON

1 SPACE PROGRAMMES around the world want to build a SHARED RESEARCH BASE ON THE MOON in the future.

2 The Moon would make an excellent STEPPING STONE for missions on their way to MARS AND BEYOND.

3 NASA plans to build a SPACE STATION TO ORBIT THE MOON.

It will act like an airport for astronauts travelling to and from the Moon.

4 There is FROZEN WATER on the Moon.

This could be extracted and converted to drinking water, oxygen for breathing, and even rocket fuel.

5 NASA has identified the BEST SPOTS FOR SOLAR PANELS to power a lunar settlement, where the Sun will shine 96 PER CENT OF THE TIME.

People will need to live in secure shelters with breathable air.

1

TUNDRA covers about 20 PER CENT of Earth's land SURFACE.

It is a vast, cold, almost treeless habitat found mostly around the Arctic.

13
August
TUNDRA

2

The GROUND is FROZEN for most of the year.

The top layer thaws in summer, but there's a constantly frozen layer under the surface, known as permafrost.

Over the year, the top "active" layer freezes and thaws.

The permafrost layer can be as much as 600 m (2,000 ft) thick.

Snowy owl

3

GLOBAL WARMING means permafrost is MELTING, which in turn SPEEDS UP global warming!

Methane gas stored in frozen soil is released when it melts, causing the atmosphere to get even warmer. This then melts more permafrost.

The musk ox has a long, double-layered, shaggy coat. In winter it digs through the snow to find grass or moss.

4

Most trees can't grow in the tundra – the FROZEN GROUND stops them putting down ROOTS.

Only small, shallow-rooted shrubs, flowers, and mosses can survive here.

The Arctic fox grows a thick white coat in winter to help it blend in with the snowy terrain.

Reindeer

5

SUMMER here lasts only a couple of months, during which there is almost 24-HOUR DAYLIGHT.

Plants burst into colourful flower, insects hatch, and 200 bird species arrive to feed and breed.

Smith's Longspur

Arctic hare

168

14
August
DENSITY

1 DENSITY is a MEASURE of HOW COMPACT the material is in an object.

2 TWO OBJECTS of the SAME SIZE can have DIFFERENT DENSITIES if they each have a different mass – different amounts of matter packed inside them.

3 A smaller object with the same mass as a larger one is DENSER because the matter is SQUASHED INTO A SMALLER VOLUME.

4 When most materials are HEATED, their particles MOVE AROUND MORE, taking up more space, so their DENSITY DECREASES.

5 When LIQUIDS of different densities are MIXED TOGETHER, they will SEPARATE so the least dense is on top and the most dense is at the bottom.

The denser liquids sink to the bottom, while the less dense liquids float to the top.

15
August
URANUS

1 Uranus is an ICE GIANT.
It was named after the Greek god of the sky, Uranos.

2 Uranus appears to SPIN ON ITS SIDE, like a rolling ball.
Scientists believe its unique tilt was caused by an object knocking into it early in its formation.

3 It is the COLDEST planet, even though it is NOT FURTHEST from the Sun.
This is because it generates little heat of its own.

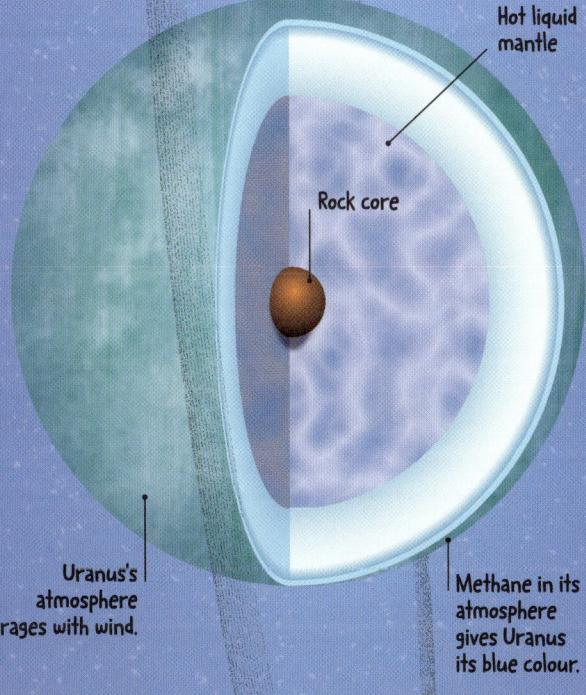

Hot liquid mantle

Rock core

Uranus's atmosphere rages with wind.

Methane in its atmosphere gives Uranus its blue colour.

4 Uranus has 13 FAINT RINGS and 28 known MOONS.
Its 13 inner moons are closely connected with its rings.

5 It takes Uranus 84 EARTH-YEARS to orbit the Sun.
That means a year on Uranus is 30,687 Earth-days long.

16
August
CRUSTACEANS

1 CRUSTACEANS are a group of animals that includes CRABS, LOBSTERS, BARNACLES, and WOODLICE.

2 A crab SHEDS its SKIN about 20 TIMES.
When it outgrows its exoskeleton, it sheds it in a process called moulting.

3 Some crustaceans REGROW damaged or lost LIMBS when they MOULT.

4 The largest crustacean, the JAPANESE SPIDER CRAB, can span 4 m (13 ft).

5 There are approximately 700 TRILLION KRILL in the Southern Ocean.
These tiny crustaceans form a vital part of the food chain.

Most crabs have two claws and eight walking legs.

Jointed limbs

Hard, rigid exoskeleton

17
August
ROCKS

1 The ROCKS in Earth's CRUST are made of combinations of different MINERAL CRYSTALS (see p180).

2 Rocks can take from ONE DAY to MILLIONS OF YEARS to form.
There are three types of rock: igneous, sedimentary, and metamorphic (see p80).

3 When MAGMA cools, IGNEOUS ROCKS form.
Large crystals form if it cools slowly, while small crystals form when it cools quickly.

Obsidian, an igneous rock

4 SEDIMENTARY ROCKS form near Earth's SURFACE.
They are made when tiny particles of rock deposited on the sea or river bed are squashed together.

Travertine, a sedimentary rock

5 METAMORPHIC ROCK forms UNDERGROUND.
It is made when rock is baked by magma or compressed by intense pressure.

Serpentinite, a metamorphic rock

1 Spacesuits PROTECT ASTRONAUTS and KEEP THEM ALIVE when they are outside a space station or walking on the Moon.

2 Spacesuits are made up of MANY LAYERS.
The tough outer layer prevents the suit from bursting or getting damaged by tiny meteoroids.

Camera

Water tank

The visor is coated with a thin layer of gold foil that protects against glare from the Sun.

3 Inside, a COOLING SUIT circulates water through 90 m (300 ft) of TUBING to remove excess body heat.

An oxygen tank supplies breathable air.

Control unit for the backpack

The gloves are heated to keep the astronaut's hands warm.

4 A PRESSURE GARMENT keeps pressure on the astronaut's SKIN.
It is designed to mimic the air pressure a person experiences on Earth.

Water circulates from the backpack and around the cooling suit.

5 A BACKPACK acts as a LIFE SUPPORT system.
It contains a water tank, oxygen tanks, a "scrubber" for absorbing carbon dioxide, a battery, a fan, and a radio transmitter.

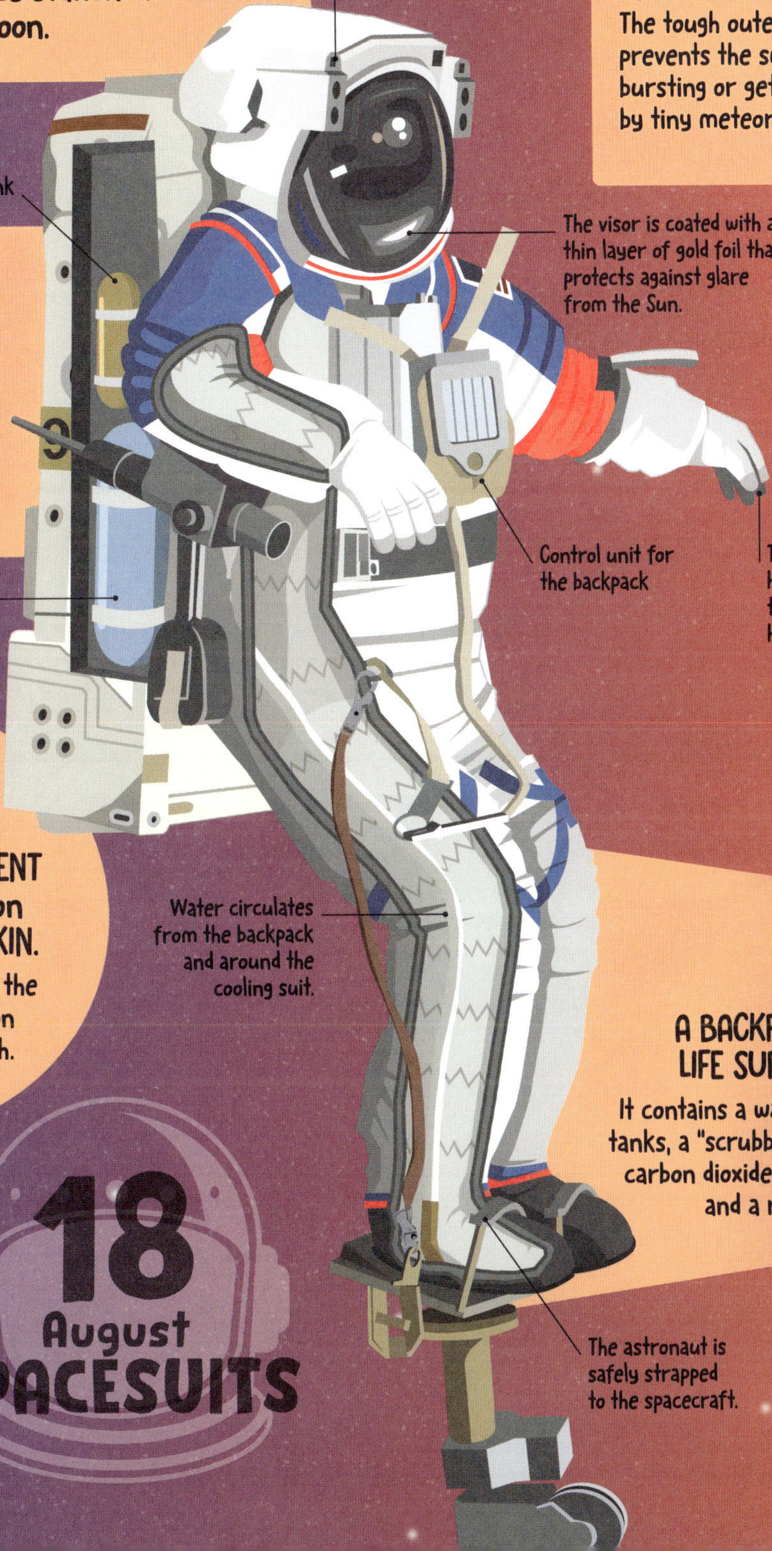

18
August
SPACESUITS

The astronaut is safely strapped to the spacecraft.

171

19
August
MOUNTAINS

1

Mountains are TALL LANDFORMS with STEEP, SLOPING SIDES.

They usually have to be more than 300 m (1,000 ft) above their surroundings to be classed as a mountain rather than a hill.

Vinicunca – the rainbow mountain – in Peru has stripes of colour across its sides.

2

They usually form when TECTONIC PLATES (see p72) COLLIDE.

These collisions cause chunks of Earth's crust to rise up.

When two tectonic plates crash into each other, the rock layers buckle upwards into a fold mountain range.

FOLD MOUNTAIN

Fault-block mountains are formed when there is movement near a crack in the crust.

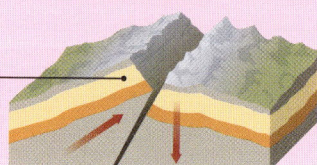

FAULT-BLOCK MOUNTAIN

Rising magma can force rock near the surface to rise up into a dome.

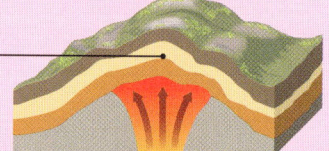

DOME MOUNTAIN

3

Mount Everest IN THE HIMALAYAS mountain range in Asia is the TALLEST MOUNTAIN IN THE WORLD.

It rises 8,848 m (29,029 ft) above sea level.

4

Some of the LONGEST MOUNTAIN RANGES are UNDERWATER. These underwater RIDGES stretch all around the globe.

5

There are MOUNTAINS ON THE MOON and on other planets.

The biggest mountain in the Solar System – Olympus Mons – is on Mars.

Olympus Mons is more than twice as tall as Mount Everest.

20
August
STORMS

2 In a storm, WINDS REACH SPEEDS of more than 75 km/h (47 mph).

Air is constantly pulled into the storm, forming swirling currents.

3 Hurricanes are a type of ROTATING STORM. They normally FORM OVER TROPICAL OCEANS and are also known as CYCLONES and TYPHOONS.

4 The biggest hurricanes can be more than 2,000 km (1,200 miles) wide!

They can cover entire countries.

1 Storms are weather events that usually include STRONG WINDS, large amounts of RAIN, and other ATMOSPHERIC DISTURBANCES.

5 It DOESN'T ALWAYS RAIN in a storm. DUST STORMS in desert climates (see p17) blow large CLOUDS OF SAND or DUST into the air.

---

1 The FIRST MACHINES TO TRANSMIT IMAGES electronically were invented in the late 1800s.

But modern televisions (TVs) didn't become available for people to buy until the 1940s.

Each tiny dot on a screen is a pixel that contains red, green, and blue elements.

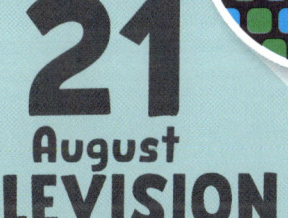

21
August
TELEVISION

3 The IMAGE YOU SEE on a TV screen is made of TINY SQUARES of light called PIXELS.

The more pixels, the higher the resolution (how sharp the picture is).

2 TVs work by RECEIVING SIGNALS and turning these into images.

Signals can be received by fibre-optic cables (see p99) or sometimes aerials mounted on top of houses.

4 TVs weren't always flat screen.

Older TVs used a tube that emitted electrons (see p53) and were big and bulky.

5 Today's HIGHEST RESOLUTION 8k TV screens contain more than 33 MILLION PIXELS!

The zombie fungus makes an ant climb up to a high place, where it dies.

A fruiting body of the fungus (see p232) erupts through the ant's head, ready to scatter its spores.

1

A PARASITE is an ORGANISM that LIVES OFF ANOTHER.

The organism that is lived off is called a host.

2

One parasite known as the "ZOMBIE FUNGUS" can take over an ant, CONTROLLING ITS MOVEMENTS!

3

The TONGUE-EATING LOUSE replaces its host FISH'S TONGUE.

The louse cuts the blood supply to the tongue, then takes its spot!

4

TAPEWORMS are parasites that live in HUMAN INTESTINES.

They can reach 15 m (50 ft) long and feed off our digested food.

22
August
PARASITES

5

The LARGEST FLOWER IN THE WORLD is a parasite.

Rafflesia plants can no longer photosynthesise (see p136), so steal food from a host vine.

23
August
G-FORCE

1 G-FORCE is a measure of the FORCE ON AN OBJECT when it is ACCELERATING.

2 You experience 1G while STANDING STILL.

If you jump off a wall, you feel zero G-force while you are falling.

3 FIGHTER and AEROBATIC PILOTS experience UP TO 9G when accelerating – that's NINE TIMES MORE than normal.

4 ASTRONAUTS wear SPECIAL SUITS that help their blood flow and stop them PASSING OUT when G-force is high.

When the carriage crests a hill or plummets quickly, you experience negative G-force and feel weightless.

When a rollercoaster turns sharply, G-force pushes you back into your seat, making you feel heavier.

5 G-force is what makes your tummy feel funny when you TAKE OFF IN A PLANE or RIDE A ROLLERCOASTER.

24
August
NERVOUS SYSTEM

The brain is the body's control centre. Together with the spinal cord, it forms the central nervous system.

1
Your nervous system is made up of your BRAIN, SPINAL CORD, and a NETWORK OF NERVES (see p10), LINKING your BRAIN to every part of your BODY.

2 Your nervous system RUNS ON ELECTRICITY.

It is very a low voltage though – just one-tenth of a torch battery.

The spinal cord is protected by the bones of your spine.

4
The tingling sensation when you hit your "FUNNY BONE" comes from hitting the ULNAR NERVE, not your ELBOW BONE.

3
REFLEX ACTIONS, such as gagging or blinking, are NOT UNDER YOUR CONSCIOUS CONTROL.

They are processed by a primitive part of the brain, and happen without you having to think about them.

Nerves are made up of bundles of nerve cells.

If you touch something hot, receptors in the fingers send a message to the brain, which then coordinates a response.

Sciatic nerve

Nerves interact with muscles. Triggering leg mucles to pull on your skeleton allows you to walk.

5 The LONGEST NEURON in your body can be MORE THAN A METRE (3 ft) long.

Called the sciatic nerve, it runs from the bottom of the spinal cord all the way to your toes.

Nerves branch out from the spinal cord to provide every part of your body, even the tips of your toes, with sensation.

1 A SOLAR ECLIPSE occurs when the MOON BLOCKS THE SUN in the sky and CASTS A SHADOW OVER EARTH.

2 Although the Sun is 400 times wider than the Moon, it is also AROUND 400 TIMES FURTHER AWAY – meaning the MOON IS ABLE TO completely BLOCK THE SUN.

Orbit of the Moon

SOLAR ECLIPSE

SUNLIGHT

EARTH

The Moon's shadow falls on Earth.

3 Sometimes the MOON AND SUN AREN'T exactly IN LINE, so only part of the Sun is blocked. THIS IS CALLED A PARTIAL ECLIPSE.

In a total eclipse, the whole Sun is blocked by the Moon and the sky goes dark.

Part of the Sun stays visible in a partial eclipse.

25
August
ECLIPSES

LUNAR ECLIPSE

SUNLIGHT

EARTH

Earth's shadow falls on the Moon.

4 A TOTAL SOLAR ECLIPSE IS SEEN somewhere on Earth EVERY 18 MONTHS.

Partial eclipses occur more often, with at least two happening each year.

5 A LUNAR ECLIPSE is when EARTH CASTS A SHADOW on the MOON.

Sunlight scattered by Earth's atmosphere still reaches the Moon, turning it reddish.

26
August
REPRODUCTION

1 REPRODUCTION is the process by which LIVING ORGANISMS CREATE MORE OF THEMSELVES.

2 In humans, REPRODUCTION begins when TWO SEX CELLS MEET.

A tiny swimming sperm cell enters the female body and meets an egg cell.

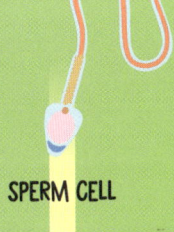

SPERM CELL

3 THEY FUSE AND SHARE DNA (see p29) in a process called FERTILIZATION.

Half of the new baby's DNA comes from the mother and half from the father.

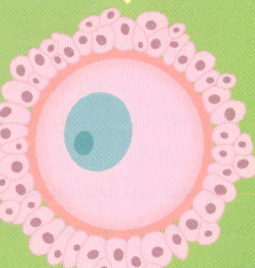

EGG CELL

4 The FUSED CELL IS called A ZYGOTE.

After 24 hours, it starts dividing. It forms a cluster of cells called an embryo. Part of the embryo will grow into a foetus then a baby.

ZYGOTE

5 It takes NINE MONTHS before a human baby is READY TO BE BORN.

This period is known as pregnancy. Pregnancy lasts 22 months in elephants, but only 3 weeks in mice!

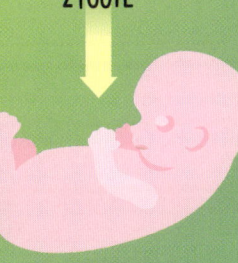

FOETUS

27
August
PENDULUMS

1 A pendulum is a WEIGHT THAT SWINGS back and forth from A FIXED POINT (a pivot).

The longer the pendulum, the slower it swings.

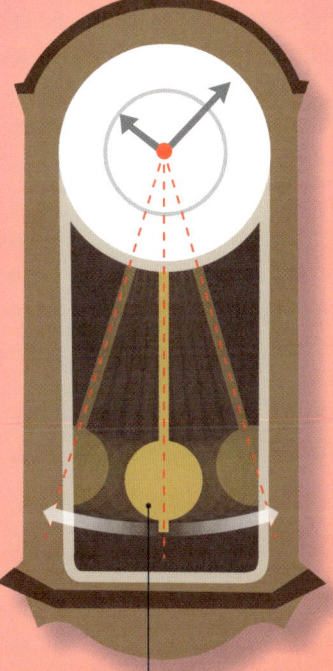

2 A pendulum ALWAYS TAKES the SAME TIME to MAKE ONE SWING.

The Italian scientist Galileo discovered this in 1583. His discovery led to the development of more accurate clocks.

3 Mechanical clocks use PENDULUMS TO KEEP TIME.

Mechanical watches use tiny, rocking wheels that work in a similar way.

Pendulum

4 MUSICIANS USE AN UPSIDE-DOWN PENDULUM CALLED A METRONOME to keep a steady beat.

The metronome mades a loud tick with each swing. Moving an adjustable weight changes the tempo.

5 The French scientist Foucault USED A PENDULUM TO PROVE EARTH ROTATES.

The direction of the swing of Foucault's 67 m (220 ft) tall pendulum gradually changes as Earth rotates.

28
August
HEARING

1

The ear is a brilliant FUNNEL for COLLECTING SOUND.

The cup-shaped part you can see is called the pinna. Once sound waves enter, they travel along the ear canal and make the eardrum vibrate.

2

The MIDDLE EAR AMPLIFIES sound.

Here the eardrum's vibrations travel through three tiny bones called the ossicles into the fluid in the inner ear.

Fluid-filled tubes called the semicircular canals help with balance.

Vibrations stimulate hairs in the cochlea and are turned into electrical signals.

The vibrations travel through the three ossicles – the smallest bones in the body.

INNER EAR

MIDDLE EAR

Signals are carried along a nerve to the brain.

OUTER EAR

Sound waves make the ear drum vibrate.

PINNA

Sound waves vibrate through the air in the ear canal.

3

Sound vibrations are converted into ELECTRICAL IMPULSES in the INNER EAR. From here they are sent to the BRAIN.

5

Your ears collect sounds, but it's your BRAIN that does the HEARING.

The brain receives the electrical signals, and uses its incredible processing power to work out what you are hearing.

4

Having TWO EARS helps us work out a sound's DIRECTION.

A sound reaches one ear a split-second before the other. This allows your brain to compare the two and work out where the sound came from.

29
August
NEPTUNE

1 Neptune is an ICE GIANT with a THICK ATMOSPHERE.

The methane in its atmosphere gives it a blue appearance.

2 It takes 165 EARTH YEARS for Neptune to make a SINGLE ORBIT of the Sun.

This is because it orbits 4.5 billion km (2.8 billion miles) from the Sun.

Neptune's liquid mantle layer weighs ten times more than Earth.

Faint rings and arcs

Atmosphere

Dark spots on Neptune's surface are huge storms.

Rocky core

There may be a layer of liquid diamond between the mantle and core.

3 It has "ARCS" and RINGS.

Neptune has several "arcs" (partial rings) made of clumps of ice and dust.

4 Its WINDS are nearly TEN TIMES FASTER than Earth's strongest HURRICANES.

5 It was the first planet to be discovered using MATHS instead of a TELESCOPE.

Scientists discovered Neptune because they noticed some unusual changes in the orbit of Uranus.

The blue-ringed octopus is both poisonous and venomous!

30
August
POISON

1 Poison is a TOXIC CHEMICAL used to DETER predators.

It is different from venom (see p13) which must be injected into the skin with a bite or sting.

2 Poison HARMS or KILLS predators when they SWALLOW, BREATHE, or ABSORB it through their skin.

The golden poison frog's bright colour warns predators it is poisonous.

3 The tiny GOLDEN POISON FROG may be the MOST POISONOUS animal on Earth.

Its skin contains enough toxin to kill 10 people!

The koppie foam grasshopper eats toxic plants to produce its poison.

4 The KOPPIE FOAM GRASSHOPPER releases poisonous FOAM to put off hungry predators.

5 Some PLANTS and FUNGI have evolved to be poisonous as a form of PROTECTION.

Just 10 berries from the deadly nightshade plant are enough to kill a human.

Deadly nightshade berry

31 August
MINERALS

1 All the ROCKS on EARTH and in SPACE are made of MINERALS.

2 MINERALS can be composed of SINGLE ELEMENTS, like silver and gold, or COMPOUNDS of elements.

Quartz is a compound of silicon and oxygen.

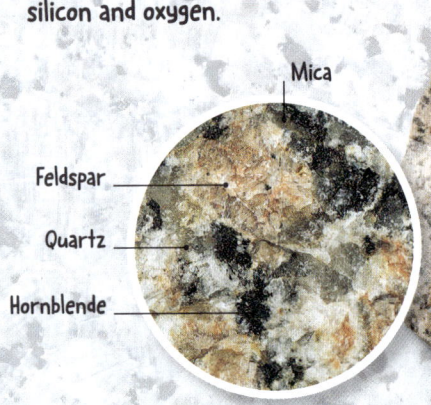

Mica

Feldspar

Quartz

Hornblende

Pink granite rock includes several different minerals.

3 Some ROCKS are made of just ONE MINERAL, but others, like pink granite, are a MIXTURE of minerals.

4 We know of more than 5,000 DIFFERENT MINERALS on EARTH.

Each one has its own chemical "recipe".

5 90 PER CENT of Earth's CRUST is made up of SILICATES – minerals composed of silicon, oxygen, and other elements.

1 September
CERAMICS

1 CERAMICS are non-metallic materials that come from NATURAL MINERALS like clay.

2 Ceramics have HIGH MELTING POINTS and are excellent HEAT INSULATORS.

Ceramic tiles are used on spacecraft to protect them from extreme heat when re-entering Earth's atmosphere.

3 Humans have been MAKING and MOULDING ceramic materials for 30,000 YEARS.

4 Ceramics are made by HEATING the ORIGINAL MATERIALS to high temperatures.

This makes them harder and stronger.

Roughly 24,300 ceramic tiles covered the surface of NASA's Space Shuttle.

5 Some ceramics can BOND with BONE, making them the perfect materials for replacement JOINTS.

1

NUCLEAR ENERGY is released by SPLITTING apart ATOMS (see p53).

This process is called nuclear fission.

2

In a NUCLEAR REACTION, particles called NEUTRONS are FIRED at the nucleus of an UNSTABLE ATOM.

The atom splits and releases neutrons, which then hit other nuclei at high speed.

A neutron is fired at the nucleus.

The nucleus splits in half and more neutrons are released.

3

A special type of URANIUM is the main FUEL used in nuclear reactions.

Its atoms easily split apart when hit by a moving neutron.

The new neutrons are fired into more nuclei, creating a chain reaction.

4

There are more than 400 NUCLEAR POWER STATIONS worldwide.

They provide around 9 per cent of all the world's electricity.

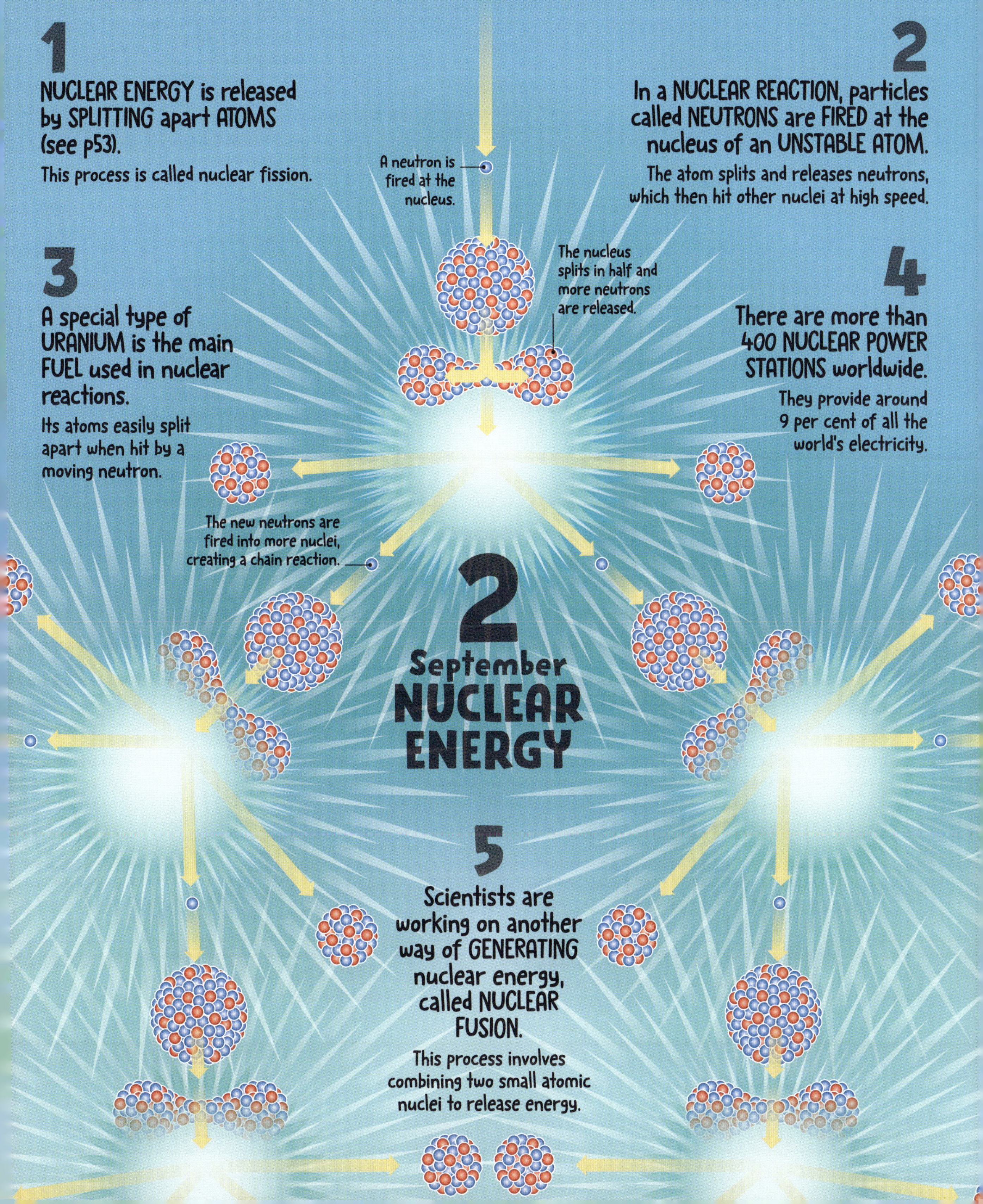

2

September
NUCLEAR ENERGY

5

Scientists are working on another way of GENERATING nuclear energy, called NUCLEAR FUSION.

This process involves combining two small atomic nuclei to release energy.

3 September
NATURAL GAS

1 NATURAL GAS IS A FOSSIL FUEL (see p212), made up of methane along with other gases.

2 Gas is often FOUND NEAR OIL DEPOSITS.

It can also form in shale rock or near coal.

3 Natural gas is naturally odourless, but AN ARTIFICIAL SCENT IS ADDED TO IT WHEN IT IS STORED.

This helps to detect any leaks.

4 FRACKING can be used TO EXTRACT NATURAL GAS.

Substances are injected into the ground, making the rock crack and release trapped gas.

5 When burned, NATURAL GAS EMITS LESS CARBON DIOXIDE and other pollutants THAN COAL or OIL.

Natural gas is cooled to form a liquid, which takes up much less space, for transportation.

4 September
HEAT

1 HEAT is the ENERGY that makes the PARTICLES IN MATTER (see p134) MOVE.

The hotter a substance is, the more its particles move.

2 There is MORE HEAT energy STORED in an ICEBERG than a HOT DRINK!

Although the particles in the drink are moving faster, the iceberg overall has a lot more particles.

3 Heat energy can be TRANSFERRED from one object to another in THREE WAYS: CONDUCTION, CONVECTION, and RADIATION.

4 BLUE or VIOLET FLAMES are the HOTTEST.

This is followed by yellow and orange flames and then red flames, which are cooler.

5 The LOWEST TEMPERATURE possible is called ABSOLUTE ZERO.

It is the point when particles stop moving, and occurs at –273°C (–459°F).

The bottom of the saucepan is in direct contact with the flame, so heats up by conduction.

Hotter parts of the water rise away from the heat source and cooler parts sink – warming up the water by convection.

Energy is transferred away from the flame through the air by radiation.

5
September
FORESTS

1

FORESTS COVER about 30 PER CENT of all EARTH'S LAND.

Russia has the most land covered by forest, followed by Brazil, then Canada.

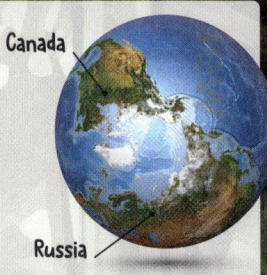

Canada

Russia

Boreal forests cover around 10 per cent of the world's land surface and spend most of the year covered in snow.

2

There are THREE MAIN TYPES: BOREAL, TEMPERATE, and TROPICAL FORESTS.

Temperate forests change with the seasons, as their trees lose their leaves in autumn.

Tropical forests, such as rainforests (see p96) are warmer and wetter than other forests.

3

BOREAL FORESTS are the MOST COMMON.

They grow in cold northerly regions and contain conifer trees with needle-like leaves.

5

Forests are crucial for TACKLING CLIMATE CHANGE.

They absorb billions of tonnes of carbon dioxide from the atmosphere every year.

4

AN AREA THE SIZE OF ICELAND is lost every year to deforestation.

Trees are cut down to make room to grow crops or rear cattle.

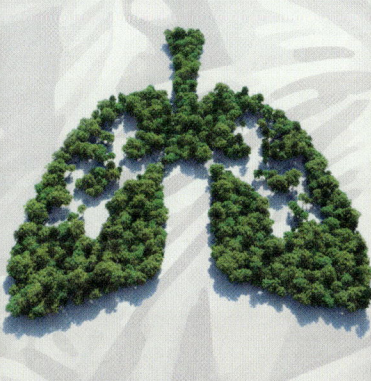

6
September
AGE-DEFYING ANIMALS

1 Jonathan the Seychelles giant tortoise is the OLDEST LIVING LAND ANIMAL.

He is estimated to be over 190 years old.

2 Scientists have observed GREENLAND SHARKS that are more than 400 YEARS OLD.

This means they may have been alive at the same time as Shakespeare!

3 The longest-living mammals are BOWHEAD WHALES, which may live over 200 years.

The bowhead whale has a gene mutation that may help it to live longer by protecting it from certain diseases like cancer.

Greenland sharks take around 150 years to reach maturity.

4 A clam called the OCEAN QUAHOG GROWS SO SLOWLY that it can live for over 500 years.

5 The OLDEST LIVING ANIMAL on Earth is thought to be a glass sponge.

It may be able to live for 15,000 years.

7
September
COMBUSTION

1 Combustion is a CHEMICAL REACTION that CAUSES SOMETHING TO BURN.

Every fire is a sign of combustion occurring.

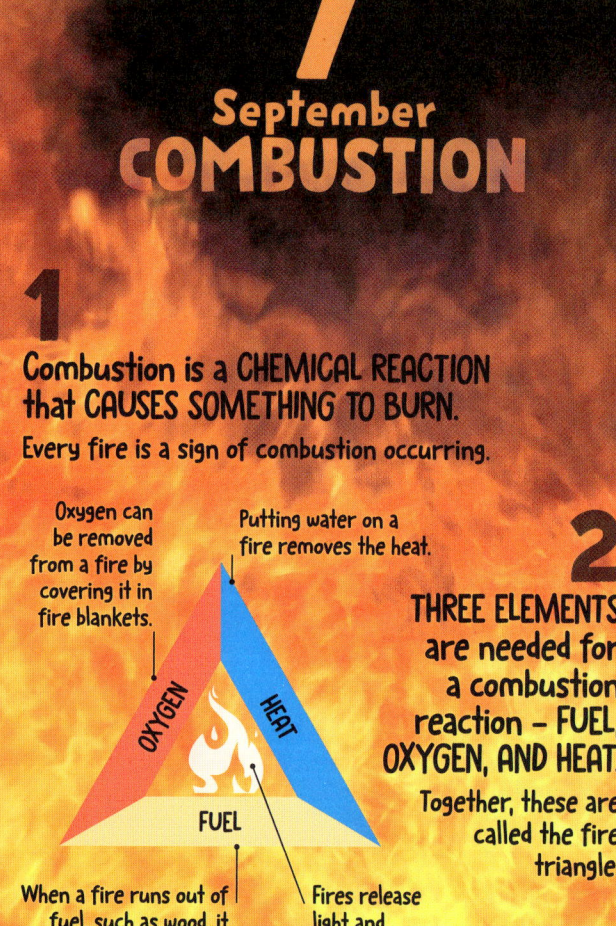

Oxygen can be removed from a fire by covering it in fire blankets.

Putting water on a fire removes the heat.

OXYGEN

HEAT

FUEL

When a fire runs out of fuel, such as wood, it will no longer burn.

Fires release light and heat energy.

2 THREE ELEMENTS are needed for a combustion reaction – FUEL, OXYGEN, AND HEAT.

Together, these are called the fire triangle.

3 To stop a fire burning, you must REMOVE ONE PART of the fire triangle.

Taking away the oxygen is called smothering.

4 THE HOTTEST FIRES can reach temperatures of MORE THAN 3,400°C (6,152°F)!

5 Before we understood combustion, SCIENTISTS USED TO THINK everything that burned contained a mysterious SUBSTANCE CALLED PHLOGISTON.

1 SIMPLE MACHINES are devices that turn SMALL FORCES into BIG ones, MOVE a force, or CHANGE THE DIRECTION of a force.

They help us to do work more easily. A wedge is a machine that helps us to split things apart, like an axe splitting wood.

A wedge turns a downward force on the thick end into a sideways force at the narrow end.

2 A LEVER is a straight bar that PIVOTS on a support, like a seesaw.

When a small force is put on one end of the lever, it is transferred through the support to the other end, making it easier to lift a load (see p201).

Pulling on the handle rotates the hammer around the head and lifts the nail.

The support is called the fulcrum.

8 September SIMPLE MACHINES

3 A PULLEY is a ROPE wrapped around a WHEEL.

It is used to transfer energy and movement. Pulling one end of the rope lifts an object attached to the other end.

Running a rope around a wheel reduces the effort of lifting an object.

4 A WHEEL can be connected to a ROD called an AXLE.

When a force rotates either the wheel or the axle, the other part also rotates. Many vehicle designs rely on wheels and axles (see p116).

An axle joins two wheels so they can move at the same time.

5 A SLOPE or RAMP is an ANGLED SURFACE and a SCREW is like a RAMP WRAPPED AROUND a ROD.

A slope makes it easier to lift something from a low surface to a higher one. A screw is useful for holding, pushing, or pulling things together.

The slope reduces the effort needed to move the box.

The screw acts like a spiral ramp, pulling it into the wood as it turns.

9
September
HYDROGEN

3 Hydrogen is ESSENTIAL FOR LIFE. It is found in WATER and in nearly EVERY MOLECULE in ALL LIVING THINGS.

4 COMBINING HYDROGEN ATOMS is called NUCLEAR FUSION.

It happens in the core of most stars and releases a huge amount of energy as light and heat.

1 HYDROGEN is the SIMPLEST CHEMICAL ELEMENT – it has just one proton and one electron (see p53).

5 Hydrogen is a PROMISING CLEAN FUEL.

It is expensive to make, but when used it produces only water as waste.

2 It is the MOST ABUNDANT CHEMICAL IN THE UNIVERSE – it makes up about three-quarters of its mass.

10
September
RAILWAYS

1 The FIRST RAILWAY LINE IN THE WORLD opened in 1825.

It connected the towns of Stockton and Darlington in the UK.

2 FUNICULAR RAILWAYS were designed to TRAVEL UP STEEP CLIFFS.

They consist of two cars that balance each other out as they travel up and down.

3 The DISTANCE between RAILS is called the GAUGE.

Countries around the world use different gauges for their tracks.

4 The TRANS-SIBERIAN RAILWAY is the LONGEST TRAIN ROUTE. It stretches 9,288 km (5,771 miles) across Russia from MOSCOW TO VLADIVOSTOK.

5 In CHONGQING, CHINA, a railway line passes directly THROUGH A BLOCK OF FLATS.

Inside, there is a station for the residents.

1 ROVERS are ROBOTS ON WHEELS.

They carry cameras and specialist equipment controlled by scientists back on Earth to explore other planets.

2 There have been SIX SUCCESSFUL PLANETARY ROVERS: Sojourner, Spirit, Opportunity, Curiosity, Perseverance, and Zhurong.

3 So far, all successful planetary rovers HAVE BEEN ON MARS.

Scientists want to explore the planet to learn about its history, including signs of life, and to explore the possibility of putting humans on Mars in the future.

11
September
PLANETARY ROVERS

The Perseverance rover has a robotic arm with a drill, a camera, and spectrometers for digging up and studying rocks.

The camera can send out laser beams to analyse what rocks are made of.

A radio antenna communicates with Earth and probes orbiting Mars.

Joints on the robotic arm help it move into any position.

4 Rovers USE SPECIAL ANTENNAS to send huge amounts of INFORMATION BACK TO EARTH – from hundreds of thousands of photos and maps to weather reports and information about the atmosphere.

5 NASA's SPIRIT AND OPPORTUNITY ROVERS (2004–2019) contained an array of EQUIPMENT FOR STUDYING ROCKS.

They confirmed that water was once present on Mars' surface.

1
FINGERPRINTS ARE REVEALED by "dusting" with a fine powder.

This sticks to the print, making it visible so that it can be picked up on clear tape and taken for analysis.

Koala fingerprints look very similar to those of humans!

2
A person's DNA CAN BE EXTRACTED from something as tiny as a HAIR or TOENAIL.

Each DNA profile can be compared to those on worldwide databases.

Everyone has unique DNA.

Drugs can stay in the bloodstream or urine for days.

Blow flies are one of the first insects to arrive at a decomposing body.

12
September
FORENSIC SCIENCE

3
BUGS can show HOW LONG AGO SOMEONE DIED.

By looking at which insects and how many of them are on a dead body, as well as what stage their larvae are at, investigators can estimate the time of death.

4
MOST accidental POISONINGS are due to COMMON HOUSEHOLD PRODUCTS.

Toxicologists can often detect what substance is involved by examining a person's blood or urine.

5
Investigators can look on your mobile phone to find out WHERE YOU HAVE BEEN.

Digital forensics analyses data from devices, such as websites accessed and geolocation data.

2

The AGE OF DINOSAURS lasted for MORE THAN 180 MILLION YEARS.

Dinosaurs evolved about 252 million years ago. The time they roamed Earth is known as the Mesozoic Era, and is split into the Triassic, Jurassic, and Cretaceous periods.

Allosaurus was a top predator in the Late Jurassic, from 152–145 million years ago.

1

Dinosaurs were PREHISTORIC REPTILES.

Unlike other reptiles, such as flying pterosaurs and ocean-dwelling plesiosaurs (see p142), dinosaurs had a hole in their hip socket that allowed them to walk upright.

Parasaurolophus was a plant-eating dinosaur that lived in herds during the Late Cretaceous period.

3

A MASS EXTINCTION at the end of the Triassic Period WIPED OUT MANY LARGE LAND ANIMALS.

This allowed the dinosaurs that survived to dominate Earth.

Pachycephalosaurus lived in the Late Cretaceous period. It ate plants, and had a thick, dome-shaped skull, which it probably used for defence.

13 September
DINOSAURS

5

NON-BIRD DINOSAURS BECAME EXTINCT 66 million years ago.

Most scientists think they were suddenly wiped out by a huge asteroid that struck Earth.

4

MODERN BIRDS ARE LIVING DINOSAURS!

Over time, some dinosaurs evolved into birds.

Palaeontologists have discovered more than 900 species of dinosaur so far, including Velociraptor.

This hunter had 56 razorlike teeth. It also had feathers, but not for flying – they probably kept it warm on cold desert nights.

Triceratops lived during the Late Cretaceous and was wiped out 66 million years ago.

1 There are around 100,000 SPECIES OF MOLLUSC.

This group of soft-bodied invertebrates includes common garden snails and slugs as well as squid and octopuses.

ANNA'S MAGNIFICENT SEA SLUG

GIANT MUSSEL

SLOW SNAIL

BLACK-MARGINED SEA SLUG

2 Most molluscs HAVE SHELLS.

The giant clam has the biggest, which can stretch to up to 1.2 m (4 ft) wide.

Tiny black eyes cover the edges of the giant blue clam.

14 September MOLLUSCS

3 PEARLS FORM WHEN AN IRRITANT, such as a grain of sand, GETS TRAPPED inside a mollusc's shells.

Natural pearls are very rare and take years to grow.

Pearls have a shimmery surface.

5 An octopus's ARMS contain their own MINI BRAINS.

Around two-thirds of its 500 million neurons (see p10) are in the arms.

4 Limpets' teeth are made of the STRONGEST KNOWN NATURAL MATERIAL.

They cover the rasping tongue, which the limpet uses to scrape algae off rocks.

EUROPEAN CHINA LIMPET

1 Tunnels can be created FOR VEHICLES to travel through, FOR MINING, STORAGE, or TO TRANSPORT sewage or water.

3 The Gotthard Base Tunnel is the LONGEST RAILWAY TUNNEL.
It runs for 57 km (35 miles) through the Alps mountain range in Switzerland.

Giant cutterheads at the front of a boring machine rotate rapidly to slice through rock in Bangalore, India.

4 The CHANNEL TUNNEL RUNS UNDER THE SEA between the UK and France – making it the longest underwater tunnel.
It took six years to build.

15
September
TUNNELS

2 Tunnels are constructed using huge TUNNEL BORING MACHINES.

5 BADGERS live in a NETWORK OF TUNNELS called a sett.
They dig these using powerful claws.

1 There are ONE BILLION nanometers in one metre.
Nanotechnology works with particles that are smaller than 100 nanometres in length, width, or height.

3 Some SUN CREAMS CONTAIN NANOPARTICLES that absorb harmful UV light (see p221) and smooth easily into your skin.

2 Carbon NANOTUBES are TINY ROLLS OF ATOMS with many potential uses.
Invented in 1991, they are strong, flexible, and conduct electricity.

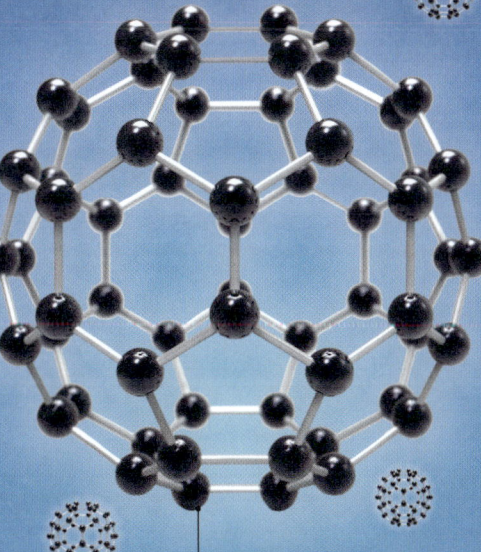

4 NANOPARTICLES can also be used on bandages TO KILL BACTERIA and to make WATERPROOF MATERIALS.

16
September
NANOTECHNOLOGY

Football-shaped buckminsterfullerene is a nanoparticle with many uses, such as in solar panels.

5 Because they are so small, NANOPARTICLES COULD BE USED INSIDE YOUR BODY to deliver drugs to specific parts!

17

September
SPREADING SEEDS

Sycamore seeds

1 To reproduce, PLANTS must SPREAD THEIR SEEDS away from the parent plant to GROW SOMEWHERE NEW.

The main ways of spreading seeds are by wind, water, animals, and explosion.

The sandbox tree's pods become "spring loaded" and eject seeds at up to 240 km/h (150 mph).

2 SOME PLANTS GROW SEED PODS THAT EXPLODE and fire seeds out in all directions.

When the seeds of the sandbox tree are ripe, they send seeds up to 100 m (330 ft) away!

3 COCONUTS FLOAT ON WATER, allowing them to drift between islands to disperse the seed inside.

Dandelion seeds

4 Seeds with FEATHERY PARACHUTES, like those in dandelion "clocks", or with PAPERY WINGS, like the seeds from sycamore trees, can catch and FLOAT ON THE BREEZE.

5 Some seeds are STICKY or HOOKED, so they CATCH ON AN ANIMAL'S FUR as it BRUSHES PAST.

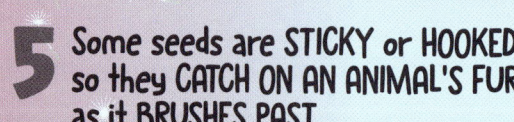

Others are tucked away inside tasty fruits that animals eat and then poo out far away.

Positively charged molecules collect at the top of the cloud.

18
September
LIGHTNING

1

Lightning happens during storms when NEGATIVE CHARGES within clouds try to connect with POSITIVE CHARGES in the ground or other clouds.

As ice particles move about inside a storm cloud, they collide and become electrically charged.

2

LIGHTNING flashes can be 30,000°C (54,000°F) – FIVE TIMES HOTTER than the surface of THE SUN.

Opposite charges between clouds are also drawn to each other.

3

Flashes of lightning move 30,000 TIMES FASTER THAN A BULLET.

Lightning occurs inside clouds when positive and negative charges connect.

4

THUNDER is the SOUND OF AIR around a lightning bolt EXPANDING SUDDENLY.

Negatively charged molecules collect at the bottom of the cloud.

5

A PARK RANGER in the US, Roy Cleveland Sullivan, was STRUCK BY LIGHTNING SEVEN TIMES and survived them all!

The negatively charged molecules try to join up with positively charged molecules on the ground.

As the negative charge moves down to the ground, a positive charge moves up to the cloud.

When the two charges meet, a big electric current shoots up to the cloud – this is the lightning that we see.

POSITIVELY CHARGED MOLECULES

19
September
COLOURS

1
The COLOURS we see are DIFFERENT WAVELENGTHS OF VISIBLE LIGHT.

White light is a mix of all the different colours.

White light is made up of all the different wavelengths of visible light.

The fruit looks orange because orange light reflects off it into our eyes.

2
When WHITE LIGHT shines on an object, NOT ALL WAVELENGTHS ARE REFLECTED off the object to your eyes – SOME ARE ABSORBED by the object.

3
The WAVELENGTHS of light that BOUNCE OFF an object are THE COLOURS YOU SEE IT AS.

When the light hits the fruit, all the wavelengths of light except orange are absorbed.

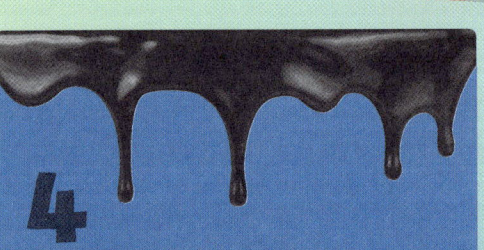

4
BLACK IS NOT A COLOUR, but an ABSENCE OF LIGHT.

Black objects do not reflect any wavelengths of light, but instead absorb almost all light.

5
BLUE is often voted to be the MOST POPULAR COLOUR, but it is one of the LEAST COMMON COLOURS FOUND IN NATURE!

20 September
MOONS

1 There are MORE THAN 250 MOONS orbiting the planets in our Solar System.

The smallest are just a few kilometres wide, but the biggest are larger than Earth's Moon.

GANYMEDE (JUPITER)

TITAN (SATURN)

CALLISTO (JUPITER)

IO (JUPITER)

THE MOON (EARTH)

EUROPA (JUPITER)

TRITON (NEPTUNE)

2 Jupiter's moon GANYMEDE is the LARGEST of all.

With a diameter of 5,270 km (3,275 miles), it is bigger than the planet Mercury!

3 Jupiter's moon IO is the MOST VOLCANIC PLACE IN THE SOLAR SYSTEM.

It has hundreds of volcanoes and lakes of molten lava on its surface.

4 MERCURY AND VENUS are the only planets that DON'T HAVE MOONS.

Jupiter and Saturn have the most, with more than 200 between them.

5 Scientists think there might be HUGE UNDERGROUND OCEANS on some of the ICY MOONS OF JUPITER AND SATURN.

They may be the places in our Solar System most likely to harbour life. Missions launched in 2023 and 2024 are heading there to find out!

21 September
ROCK POOLS

1 ROCK POOLS are found on ROCKY SHORES.

They lie in the intertidal zone – the area along shorelines that is covered by water at high tide, but exposed at low tide.

2 They are an EXTREME ENVIRONMENT.

Conditions change as the tide goes in and out, repeatedly covering then exposing the pool.

3 The BLENNY is one of the FEW FISH TO LIVE IN ROCK POOLS.

As long as it stays moist, it can survive out of water by breathing through its skin and gills.

4 Each pool is its own TINY ECOSYSTEM.

They may provide a home for seaweeds and other algae, and small creatures like barnacles and crabs.

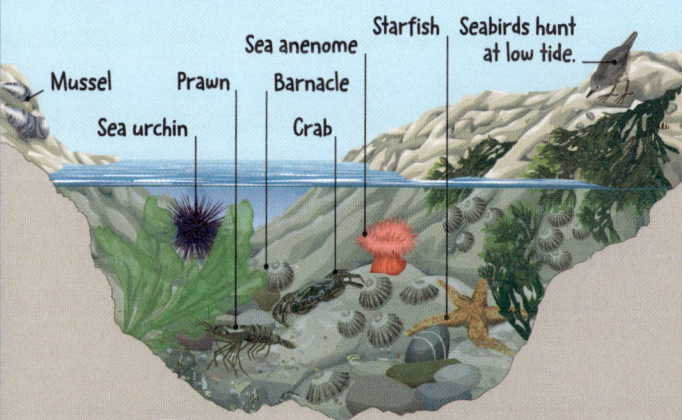

Mussel

Sea urchin

Prawn

Sea anenome

Barnacle

Crab

Starfish

Seabirds hunt at low tide.

5 SEA ANEMONES may look like plants, but they ARE ACTUALLY ANIMALS.

They attach to rocks and catch food using tentacles.

1

Danyang–Kunshan Grand Bridge in China IS THE LONGEST BRIDGE IN THE WORLD.

It stretches 164 km (102 miles) – roughly the length of 37,500 cars back-to-back!

2 In Meghalaya, India, BRIDGES HAVE BEEN MADE FROM LIVING TREE ROOTS through decades of weaving and training.

22
September
AMAZING BRIDGES

3 SOME BRIDGES ARE not designed for vehicles, but FOR WILD ANIMALS.

"Green bridges" connect green spaces and are planted with vegetation to allow wildlife to cross busy roads and train tracks more safely.

5

The Millau Viaduct in France is THE WORLD'S TALLEST BRIDGE.

It reaches 343 m (1,125 ft) from the top of its masts to the bottom of the valley below – that's higher than the Eiffel Tower.

4 Nicknamed the "Flying Drawbridge", the Slauerhoff Bridge in the Netherlands has ARMS THAT SWING a section of the bridge 90 DEGREES INTO THE AIR, SO SHIPS CAN PASS.

23 September — OIL

1 OIL IS A FOSSIL FUEL (see p212) that is often refined to make petrol, diesel, and other fuels.

2 The BIGGEST OIL RESERVE is Ghawar Field in Saudi Arabia.

It is estimated to contain around 48 billion barrels of oil.

A nodding donkey (pumpjack) pumps oil from the ground.

3 OIL RESERVES often LIE DEEP BELOW LAND OR THE SEABED.

The world's deepest oil well extends more than 12 km (7.5 miles) into Earth's crust, making it 15 times taller than the world's tallest skyscraper.

4 Geologists find oil by sending VIBRATIONS INTO EARTH.

These are reflected back differently by different rock layers, allowing geologists to work out where oil might be.

An oil well is made by drilling straight down through solid rock.

5 PLASTIC BOTTLES, BICYCLE TYRES, PAINT, SOAP, CLOTHES, and COSMETICS are some of the many products MADE FROM OIL.

OIL RESERVE

24 September — EARLY HUMANS

1 We share 98.8 PER CENT OF OUR DNA WITH CHIMPS, our closest relatives.

Our ancestors split from the ancestors of chimps 7 million years ago in Africa.

2 Our human ancestors started WALKING UPRIGHT at least 4 MILLION YEARS AGO.

This freed their hands to use tools. But they still slept in trees at night like chimps do.

Walking upright allowed our ancestors to leave the trees and forage on grasslands.

3 Early humans LOST THEIR FUR and became naked 1.5–2 MILLION YEARS AGO.

They didn't start wearing clothes until more than a million years later.

4 By at least 800,000 YEARS AGO, early humans had LEARNED HOW TO MAKE FIRE.

Using it to cook food made tough plant roots and decaying meat edible.

5 MODERN HUMANS, *Homo sapiens*, appeared roughly 200,000 YEARS AGO.

We evolved in Africa and then spread out all over the world.

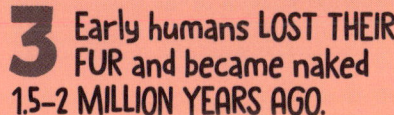

25
September
IRON

1 **EARTH IS 35 PER CENT IRON.**
Iron is the most abundant chemical element in the world, and forms most of our planet's core.

2 **BLOOD IS RED BECAUSE IT CONTAINS IRON.**
Iron atoms in the protein haemoglobin carry oxygen around the body (see p122).

There's enough iron in your body to make a small nail.

3 **The first iron that people used CAME FROM METEORITES.**
The Egyptians used meteoritic iron to make beads and daggers.

4 **People discovered how to use FURNACES TO EXTRACT IRON from ROCK around 4,000 YEARS AGO.**
This triggered the Iron Age, when iron replaced bronze in weapons and tools.

5 **Iron is the world's MOST WIDELY USED METAL.**
In the form of steel (see p221), it is used to make everything from buildings and bridges to cars and cutlery.

26
September
VERTEBRATES

1 **A vertebrate is AN ANIMAL WITH A BACKBONE.**
There are five main types: mammals, birds, amphibians, reptiles, and fish.

2 **A backbone ISN'T ONE BONE. It's a STACK of SMALL BONES called VERTEBRAE.**
The human backbone has 33 vertebrae, but snake backbones have up to 600.

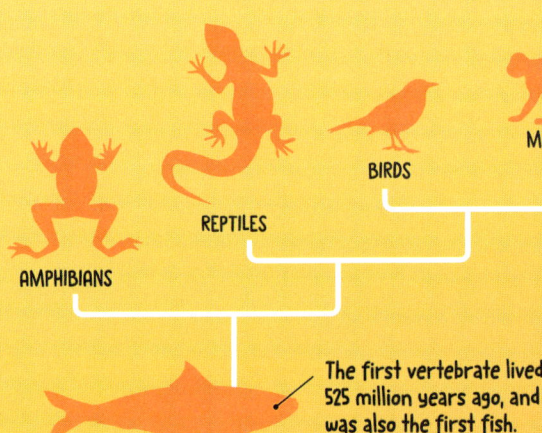

MAMMALS

BIRDS

REPTILES

AMPHIBIANS

FISH

The first vertebrate lived 525 million years ago, and was also the first fish.

3 **WE EVOLVED FROM FISH.**
All land-living vertebrates, including us, evolved from fish that clambered onto land. Their fins turned into four limbs ending in digits (fingers and toes).

4 **NEARLY ALL LARGE ANIMALS ARE VERTEBRATES.**
The largest land animals ever – the dinosaurs – were vertebrates. The largest sea creatures are vertebrates too.

5 **Almost every vertebrate HAS A TAIL.**
Humans have lost theirs, and frogs have tails only when they are tadpoles.

27 September
LIVING IN SPACE

1

PEOPLE LIVING IN SPACE must DEAL WITH MICROGRAVITY. This is when people and objects float about as though they weigh nothing.

Microgravity is caused by the high-speed orbits of spacecraft flying around Earth.

Astronauts float about in microgravity.

Liquids don't flow in space, so astronauts suck drinks from pouches.

2

ASTRONAUTS ON THE ISS (see p206) see 16 SUNRISES and 16 SUNSETS every 24 HOURS as they circle Earth.

To help them sleep, the lights inside are dimmed to simulate normal night-times.

3

TOILETS DON'T FLUSH in space.

Astronauts have to use suction tubes to hoover everything up. This can get messy, so there are lots of cleaning products next to the loo.

There are handholds everywhere to make moving easier.

Astronauts strap themselves in place to sleep. They can't lie down as there's no "down" in space.

4

In 2024, Russian cosmonaut Oleg Kononenko became the first person to spend MORE THAN 1,000 TOTAL DAYS IN SPACE!

5 Spending a lot of TIME IN SPACE WEAKENS YOUR BONES AND MUSCLES. The risk of some diseases also increases due to exposure to radiation from the Sun and space.

28
September
THE MILKY WAY

1 The MILKY WAY is ONE OF BILLIONS OF GALAXIES in the Universe. Earth and our Solar System are located in one of its SPIRAL ARMS.

2 It is a BARRED SPIRAL GALAXY
A flat disc made up of spiralling arms orbits a central bulge (see p28).

3 There are more than 100 BILLION STARS IN THE MILKY WAY.

The central bulge is densely packed with stars.

Scutum–Centaurus Arm

Clusters of old stars are found in the stellar halo.

Our Solar System is in the Orion Arm.

Outer Arm

Perseus Arm

The spiral arms contain gas and dust that form new stars (see p155).

4 A SUPERMASSIVE BLACK HOLE lies at the centre.
The black hole (see p41), called Sagittarius A*, has a mass four million times greater than our Sun.

5 The Milky Way is on a COLLISION COURSE with its nearest NEIGHBOUR.
In about 4.5 billion years, the Milky Way will crash into the Andromeda Galaxy.

29 September — LEVERS

1 LEVERS ARE SIMPLE MACHINES (see p185) that can MAGNIFY the effect of A FORCE.

They make it easier to lift, move, cut, or crush something.

2 When you PUSH A LEVER, the lever ROTATES around a fixed point called a FULCRUM.

The other end of the lever swings the opposite way to your hands.

3 If the force you apply (called the EFFORT), is FURTHER FROM THE FULCRUM than the load, the LEVER MAGNIFIES ITS EFFECT.

This is how a pair of pliers, scissors, a crowbar, and a wheelbarrow work.

4 If the EFFORT is NEARER to the fulcrum than the load, the LEVER REDUCES THE FORCE.

This is how tweezers work.

5 The TINY BONES in the HUMAN EAR are LEVERS.

They amplify sounds picked up by the eardrum.

EFFORT

LOAD

FULCRUM

30 September — GASES

1 A GAS is a STATE OF MATTER (see p134) where the PARTICLES are very FAR APART and CONSTANTLY MOVING around.

2 Many gases are INVISIBLE!

But others have colours: iodine gas is purple and chlorine gas is green-yellow.

3 When HEATED, gases EXPAND.

This is what makes hot-air balloons rise up into the air (see p141).

4 Unlike solids and liquids, GASES CAN BE EASILY COMPRESSED.

Their particles can be forced closer together, making them take up a smaller area.

5 FIZZY DRINKS have BUBBLES OF CARBON DIOXIDE GAS in them.

The hissing sound when you open a drink is some of this gas escaping into the air.

1

Evolution is the process of HOW LIVING THINGS CHANGE over millions of years.

These changes take place over many generations and lead to the formation of new species.

1

October

EVOLUTION

2

Animals today LOOK DIFFERENT to their prehistoric ANCESTORS because of evolution.

The ancestors of elephants had shorter trunks.

Separate finch species all had a common ancestor

Darwin noticed that the species of Galápagos finches on different islands had different beak shapes.

3

As organisms reproduce, new MUTATIONS EMERGE in their DNA (see p29), so their offspring have many DIFFERENT CHARACTERISTICS.

Each beak is adapted to a specific diet.

5

English scientists CHARLES DARWIN and ALFRED RUSSEL WALLACE came up with the theory of EVOLUTION.

The two men, who were friends, developed their ideas separately before realizing they had both made the same breakthrough.

Darwin visited the Galápagos Islands in 1835 and studied the wildlife there.

4

Characteristics are PASSED ON when they give an organism AN ADVANTAGE IN A SPECIFIC ENVIRONMENT or against a competitor.

2 October
FLUORESCENCE

1 Fluorescent objects GLOW UNDER UV LIGHT.

This happens because the UV light (see p221) is absorbed and then emitted at visible wavelengths.

2 It is named after the mineral FLUORITE, whose CRYSTALS GLOW blue under UV light.

3 Lots of FLOWERS FLUORESCE.

Scientists think this might help guide pollinating insects, which can see more UV wavelengths than humans can.

5 All SCORPIONS FLUORESCE under UV light. No one knows why!

4 Fluorescence is used to COMBAT COUNTERFEITING.

Many bank notes have features printed in UV ink so that they show up only under a UV lamp.

Fluorite crystals on quartz

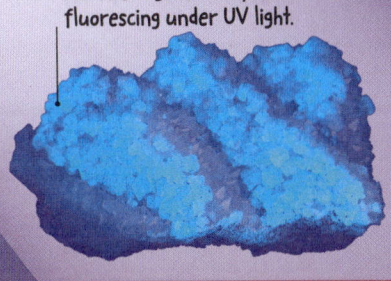

Fluorite crystals on quartz fluorescing under UV light.

---

3 October
COAL

1 COAL is a FOSSIL FUEL (see p212). It is largely made up of the element CARBON.

2 Ancient PLANT MATTER formed coal.

Dead plants turned into peat bogs that over time were transformed into coal.

3 People have MINED COAL near the surface for THOUSANDS OF YEARS, but most coal is buried deeper.

Today's deepest coal mine extends underground for 1,546 m (5,072 ft).

4 Coal was an IMPORTANT FUEL in the INDUSTRIAL REVOLUTION (1750–1900).

More mines opened to power factories and steam engines.

5 Anthracite is the HARDEST FORM OF COAL – producing a very HOT FLAME when burned.

Other types of coal contain less carbon, but are still widely used.

1

The LIGHT WE SEE is just ONE SMALL PART of a range of electromagnetic radiation known as THE ELECTROMAGNETIC SPECTRUM.

2

Electromagnetic radiation consists of WAVES THAT TRANSFER ENERGY through the Universe.

Electromagnetic waves can have different wavelengths. Their properties depend on the wavelength (see p217).

3

Scientists DIVIDE THE ELECTROMAGNETIC SPECTRUM INTO SEVEN GROUPS: gamma rays, X-rays, ultraviolet, visible light, infrared, microwaves, and radio waves.

RADIO WAVES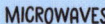

MICROWAVES

INFRARED

VISIBLE LIGHT

ULTRAVIOLET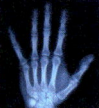

X-RAYS

GAMMA RAYS

Radio waves (see p107) have the longest wavelength and lowest energy.

Microwaves (see p229) can be used for communication as well as for cooking.

Infrared radiation (see p256) is also known as heat radiation.

Ultraviolet light (see p221) can be seen by some animals.

X-rays (see p40) can be harmful to humans in high doses.

Gamma rays are emitted by radioactive substances. They can penetrate through many materials, but cannot pass through lead.

4

October

ELECTROMAGNETIC SPECTRUM

Some flowers have patterns visible only in UV light, to attract pollinating insects that see in this wavelength.

4 Most of the electromagnetic spectrum is INVISIBLE TO US.

We can see only visible light.

5

GAMMA RAYS have the shortest wavelengths and HIGHEST ENERGY. They can be used to DESTROY CANCER CELLS in humans.

5

October
CARS

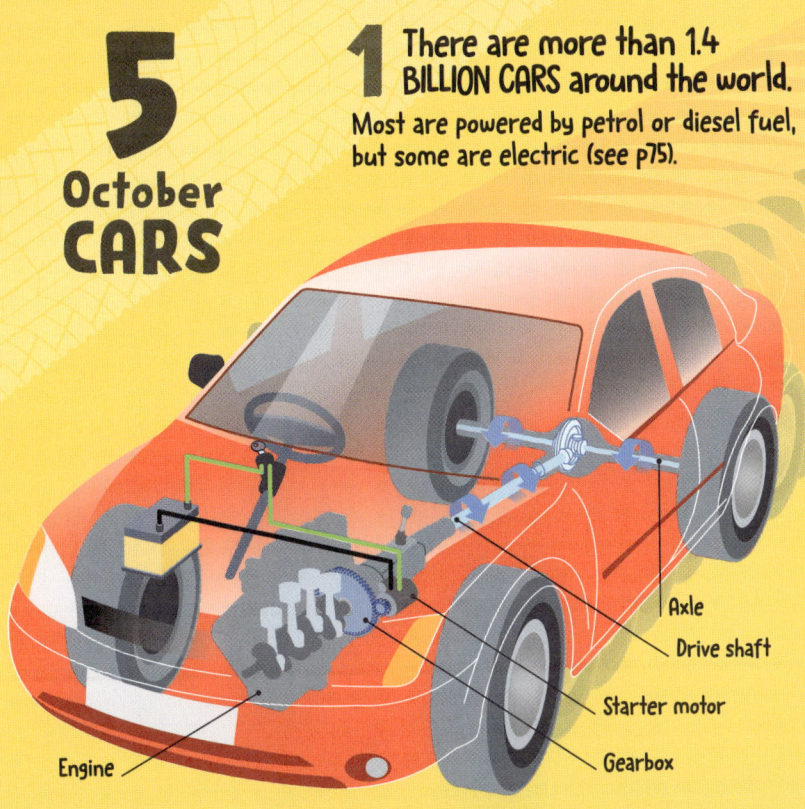

1 There are more than 1.4 BILLION CARS around the world.

Most are powered by petrol or diesel fuel, but some are electric (see p75).

2 The FIRST CAR for sale was the BENZ MOTORWAGEN, released in 1886.

It could only reach speeds of less than 16 km/h (10 mph).

3 To start a car, the driver usually PUTS THE KEY IN THE IGNITION.

This completes an electric circuit, allowing the battery to power the starter motor.

4 The motor starts the engine (see p126), which BURNS FUEL TO PRODUCE A FORCE that passes to the axle and TURNS the WHEELS.

5 A car's GEARBOX allows the force from the engine TO BE USED MOST EFFECTIVELY.

High gears are used for higher speed and low gears are used for more power.

Axle

Drive shaft

Starter motor

Gearbox

Engine

6

October
WIND

The spin of Earth on its axis affects the direction of some winds.

Winds circulate constantly around the globe.

1 Wind is the MOVEMENT OF AIR, CAUSED BY THE SUN unevenly heating Earth.

2 AT THE EQUATOR, the Sun WARMS EARTH MORE.

This warm air, which rises, moves to the poles, cools, and then flows back.

Without these circulating winds, the poles would become colder and colder.

Equatorial regions would become hotter without wind.

3 Sea breezes happen because AIR ABOVE LAND HEATS UP MORE QUICKLY than above seas, so cooler air gets pulled inland.

4 The BEAUFORT WIND SCALE measures the speed and severity of winds.

The highest level is that of a hurricane (see p173).

5 COMMONWEALTH BAY in Antarctica is often considered the WINDIEST PLACE IN THE WORLD.

Wind speeds there regularly exceed 240 km/h (150 mph).

7
October
INTERNATIONAL SPACE STATION

1 The INTERNATIONAL SPACE STATION (ISS) is an ORBITING STATION built and maintained by FIVE SPACE AGENCIES from around the world.

Solar panels capture sunlight to provide power to the ISS.

2 The station STRETCHES FOR 109 m (356 ft). A large part of this length is panels, made up of the 262,400 solar cells (see p45) that power the ISS.

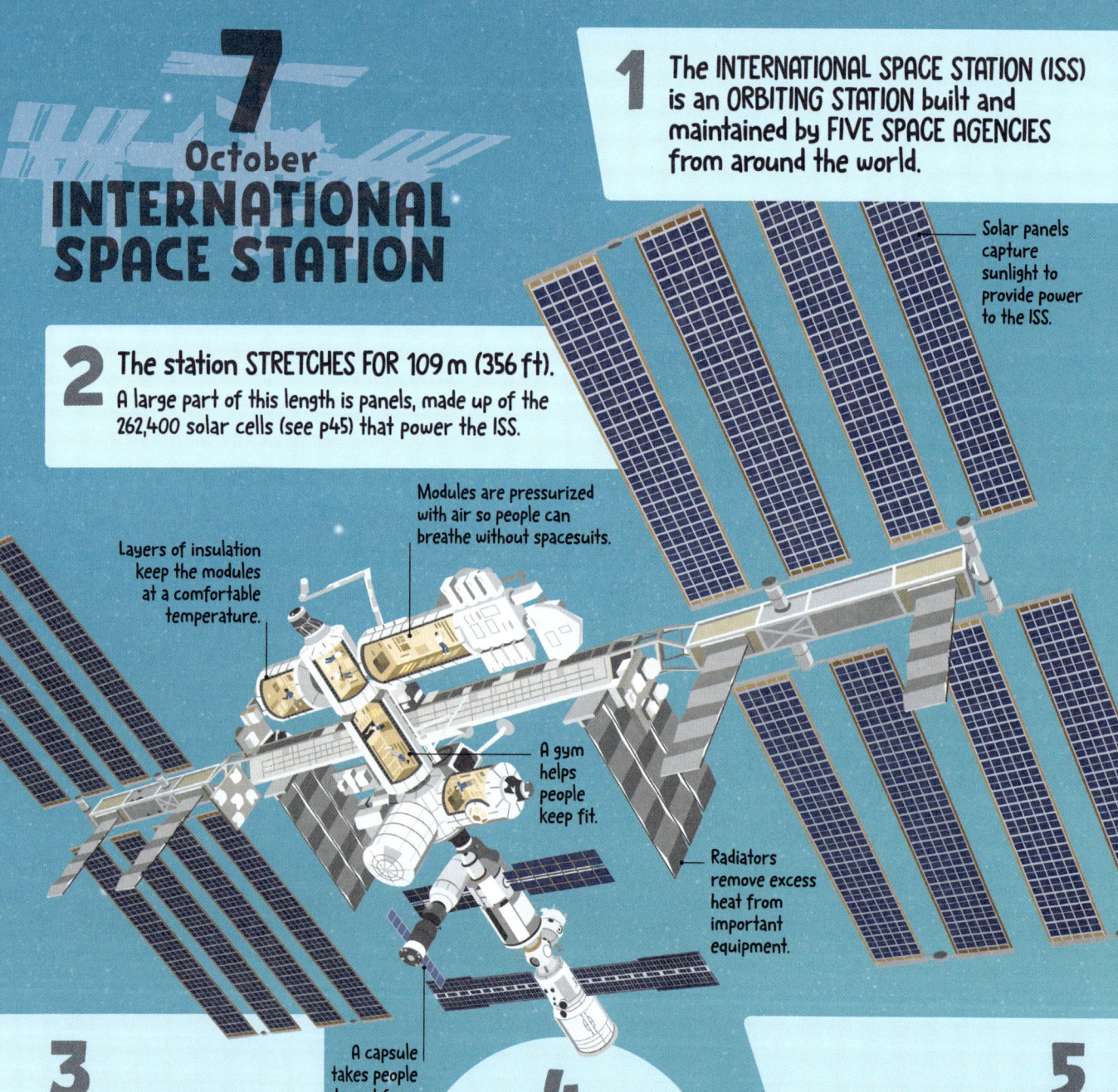

Modules are pressurized with air so people can breathe without spacesuits.

Layers of insulation keep the modules at a comfortable temperature.

A gym helps people keep fit.

Radiators remove excess heat from important equipment.

A capsule takes people to and from the ISS.

3 The FIRST PIECE of the ISS was SENT INTO SPACE IN 1998.

New modules have been added as recently as 2021.

4 More than 270 ASTRONAUTS HAVE VISITED the ISS.

It contains six sleeping quarters, two bathrooms, and a gym.

5 Scientists on the ISS study the effects of LOW GRAVITY on humans and other organisms, and explore WAYS WE COULD LIVE IN SPACE.

8 October ROBOTS

Multiple joints provide robotic arms with freedom of movement.

3 Robots can be used for SEARCH AND RESCUE, EXPLORATION, GOODS DELIVERY, CLEANING, and even COMPANIONSHIP.

4 HUMANOID ROBOTS are designed to MIMIC HUMAN BEHAVIOUR AND MOVEMENTS, but they don't look very human – yet!

1 A ROBOT is a MACHINE THAT CAN PERFORM COMPLEX TASKS, and generally work AUTOMATICALLY.

2 SENSORS allow robots to RECEIVE INFORMATION from the world around them.

They can also use artificial intelligence (see p94) to process data they receive and make decisions.

5 Today, most robots are USED IN FACTORIES.

More than 4 million industrial robots are currently in use worldwide.

In car factories, several robots work together to build a car.

9 October WATER WAVES

1 WAVES are caused by WIND BLOWING ACROSS THE SURFACE of the open ocean.

They begin as gentle ripples, but grow larger as they roll towards the shore. Stronger winds create larger waves.

Wind makes the water surface start to move up and down as ripples.

3 WAVES TRANSFER ENERGY, not water, across the ocean.

That's why an object on the water will bob up and down in one place rather than be carried along.

4 The BIGGEST WAVE EVER SURFED measured 26.2 m (86 ft) – about the size of a SIX-STOREY BUILDING.

ENERGY

2 Waves make water move in a CIRCULAR MOTION.

As the water rolls forwards, it transfers energy to water in front.

The water moves in circles, transferring energy forwards.

5 Sometimes, TWO WAVES WILL JOIN to form huge ROGUE WAVES.

They are unpredictable and can sink ships and damage oil rigs.

1

A RENEWABLE source of energy is ONE THAT CANNOT BE USED UP, unlike fossil fuels (see p212).

Wind, water, and the Sun are renewable energy sources.

2

Renewable energy sources usually HAVE LITTLE or NO HARMFUL IMPACT ON THE ENVIRONMENT.

3

HYDROPOWER (see p109) is the type of renewable energy that has been USED THE LONGEST.

Water wheels were used in many ancient cultures.

Wind pushes turbines (see p76) around to generate energy.

4

MORE THAN 30 PER CENT of the world's ELECTRICITY comes from RENEWABLES – up from 19 per cent in the year 2000.

Solar panels (see p45) can convert light energy from the Sun into electricity.

Hydropower uses flowing water to turn turbines.

Harnessing the power of tides or natural waves is another way of using water to generate energy.

Geothermal energy (see p124) uses the heat from within Earth.

10
October
RENEWABLE ENERGY

5

SCIENTISTS are DEVELOPING ways to use NEW RENEWABLE SOURCES.

These include algae, waste body heat, and the energy of people pushing on pavements.

11
October
RAINBOWS

1 RAINBOWS are an OPTICAL ILLUSION.
They appear when light from the Sun passes through raindrops and splits into different colours.

2 TO SEE A RAINBOW, the SUN must be BEHIND YOU and the RAIN IN FRONT OF YOU.
No two people will see exactly the same rainbow because the effect depends on exactly where the viewer is.

3 A full rainbow forms a circle not an arch, but SOME OF THE CIRCLE IS HIDDEN BELOW THE HORIZON.
From an aeroplane, it is possible to see the full circle.

4 MOONBOWS are a rare sight that form in the same way as RAINBOWS, but WITH LIGHT FROM THE MOON rather than the Sun.

5 A DOUBLE RAINBOW appears when light BOUNCES TWICE INSIDE THE SAME RAINDROP.
The sequence of colours in the second bow are in reverse.

---

1 The Wright Flyer was THE FIRST PLANE to make a POWERED FLIGHT in 1903.
It flew for just 12 seconds.

2 Pilots fly planes by using a THROTTLE to CONTROL HOW MUCH FUEL flows TO THE ENGINE.

3 THE BIGGEST PLANES are designed to TRANSPORT SPACECRAFT, rather than humans.
The Scaled Composites Stratolaunch carries rockets that launch from the air.

The Wright Flyer's wings were made from a wooden frame covered with cloth.

4 Some aircraft can FLY FASTER THAN SOUND.
Concorde, which flew from 1976 to 2003, was the only commercial plane to do this.

5 MORE THAN 8,000 PLANES are IN THE SKY at any moment!

12
October
PLANES

1

We don't know HOW or WHEN the Universe will END, but it probably won't occur for many BILLIONS OF YEARS.

There are four popular theories for how it will end.

THE BIG BOUNCE

A new Big Bang occurs, creating a new Universe.

The Universe collapses into a huge black hole.

THE BIG CRUNCH

2

Some scientists believe the Universe will STOP EXPANDING and will CONTRACT instead in a "BIG CRUNCH".

Some think another Big Bang (see p23) could then occur, called the Big Bounce.

3

The BIG CHILL theory suggests that the Universe will EXPAND so much that it COOLS.

All stars and galaxies would die and temperatures would plummet to absolute zero.

The Universe starts to contract and galaxies merge together.

The Universe stops expanding and the Milky Way dies.

4

The BIG RIP theory proposes that the Universe could EXPAND SO FAST that it TEARS ITSELF APART.

Galaxies use up their gases, and dying stars are not replaced by new ones.

Older stars lie at the centres of galaxies.

13

October

END OF THE UNIVERSE

Milky Way

Star birth occurs in the spiral arms.

5

The BIG CHANGE is a theory that suggests a NEW, ALTERNATIVE UNIVERSE will form.

The Universe as we know it would end.

African bush elephants don't reach full size until they're 35-40 years old.

14
October
LARGEST ANIMALS

The ostrich has the largest eye of any land animal at 5 cm (2 in).

1

The **AFRICAN BUSH ELEPHANT** is the **LARGEST ANIMAL** on land.

Adult males reach 3 m (10 ft) tall and weigh up to 6,000 kg (13,000 lb).

2

The **LARGEST BIRD** on Earth is the **OSTRICH**, which can be 2.75 m (9 ft) tall.

The Chinese giant salamander absorbs oxygen through its skin!

3

The **LARGEST AMPHIBIAN** on Earth, the **CHINESE GIANT SALAMANDER**, grows up to 1.8 m (6 ft) long.

4

The **BLUE WHALE** is the **LARGEST ANIMAL EVER TO HAVE LIVED.**

At 180,000 kg (400,000 lb), it weighs more than 30 adult African elephants.

The blue whale is larger than any dinosaur.

5

The **WHALE SHARK** is the **LARGEST FISH** in the oceans, reaching an average of 12 m (40 ft) long.

Whale sharks are not a type of whale, but a species of shark.

15
October
HABITATS

1

A habitat is the **NATURAL HOME** of an animal, plant, or other living thing.

There are hundreds of different habitats. Groups of similar habitats are called biomes.

Even ice is a habitat for some creatures, like these methane ice worms.

2

A habitat provides the organisms that live there with the four **ESSENTIALS** for life: **WATER, FOOD, SHELTER,** and enough **SPACE** to survive.

3

Many animals **ADAPT** to fit their habitat, helping them to **SURVIVE.**

4

A habitat can be an animal's **PERMANENT HOME,** or somewhere it goes **TEMPORARILY** to find food, breed, or raise young.

5

Almost **EVERY PLACE ON EARTH** – from a volcanic lava field to deep inside a glacier – is a **HABITAT** for some kind of organism.

The Indonesian maleo bird uses the warmth of sun-warmed sand or the soil near hot springs or to incubate its eggs.

16
October
DIVING

1 The most successful FREEDIVERS can dive to depths of MORE THAN 150 m (490 ft) WITHOUT OXYGEN!

2 The deeper you dive in water, the GREATER the PRESSURE.

This is the force of all the water above and around pushing in on you.

A diver's tank often contains a mix of gases called nitrox.

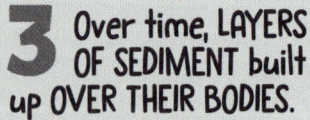

3 In SCUBA diving, DIVERS CARRY tanks of COMPRESSED AIR in an apparatus called AN AQUALUNG.

Each tank lasts for around an hour.

4 Divers wear weights and an inflatable bag TO CONTROL THEIR BUOYANCY (see p9).

They can release the air in the bag to sink.

5 When ascending back to the surface, DIVERS MUST GO SLOWLY TO AVOID a condition called "THE BENDS".

This is where the change in pressure causes nitrogen to form gas bubbles in the bloodstream. It can be deadly!

17
October
FOSSIL FUELS

1 FOSSIL FUELS ARE natural materials that have FORMED OVER MILLIONS of YEARS.

They are coal, oil, and natural gas.

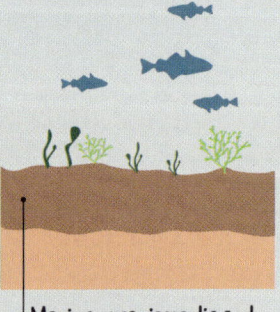

Marine organisms die and are buried under sediment.

2 Oil and gas formed FROM ANCIENT ORGANISMS that died and became buried on the seabed.

3 Over time, LAYERS OF SEDIMENT built up OVER THEIR BODIES.

The combined pressure and heat turn these into oil and gas.

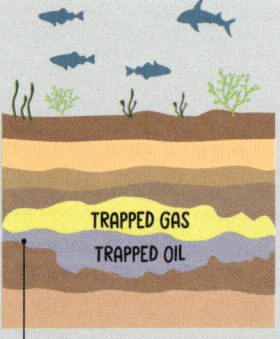

TRAPPED GAS
TRAPPED OIL

Pressure turns the remains to gas and oil.

4 BURNING FOSSIL FUELS releases CARBON DIOXIDE and OTHER GREENHOUSE GASES (see p38) – contributing to global warming.

Pipe is drilled through layers of rock to extract oil and gas.

5 It is estimated we will RUN OUT OF FOSSIL FUELS by 2100.

Existing deposits of the fuels are being used up rapidly.

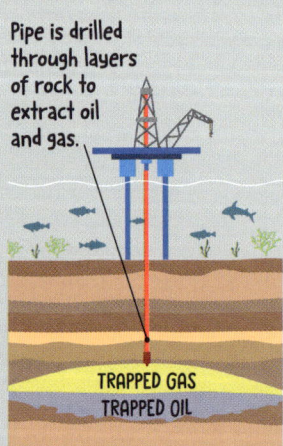

TRAPPED GAS
TRAPPED OIL

1

There are SIX ESSENTIAL NUTRIENTS our body needs: carbohydrates, proteins, fats, vitamins, minerals, and water.

18
October
FOOD

2

PROTEINS are USED to help body cells (see p214) GROW and REPAIR.

Meats, fish, dairy, and nuts are all good sources of protein.

3

CARBOHYDRATES – found in foods such as bread and rice – are the body's main SOURCE OF ENERGY.

5

Your body needs many DIFFERENT VITAMINS and A VARIETY OF MINERALS to work normally.

Eating a varied diet with lots of different foods is the best way to get all these.

4

In order for food to pass smoothly through our digestive system, we also NEED TO CONSUME ENOUGH FIBRE – a type of carbohydrate.

213

19
October
HISTORY OF MEASUREMENT

The ancient Egyptian pyramids were built using cubits.

1 Many ANCIENT UNITS of measurement were BASED ON BODY PARTS.

For example, the cubit was based on the length from the elbow to the tip of the middle finger.

2 The term "MOMENT" comes from the 8th century – a monk called St Bede USED THE WORD to describe a time PERIOD OF 90 SECONDS.

3 The length of a metre HASN'T ALWAYS BEEN the SAME.

Originally based on a fraction of Earth's circumference, it is now based on the speed of light in a vacuum.

4 The SMALLEST unit of LENGTH is the YOCTOMETRE.

There are one septillion (1 followed by 24 zeros) yoctometres in one metre.

5 A LIGHTYEAR IS THE DISTANCE THAT LIGHT CAN TRAVEL IN ONE YEAR – nearly 9.5 trillion km (6 trillion miles).

20
October
HUMAN CELLS

Axon terminals at the end of the cell pass signals to other cells.

Nerve cell (see p10)

1 There are about 200 different TYPES OF CELL IN THE HUMAN BODY, each with its own job to do.

Inside organs, cells of the same type work together in groups called tissues.

2 WE ALL GROW FROM A SINGLE CELL called AN OVUM (egg).

This joins with a sperm cell then divides and multiplies to create all the different cells in the body.

Muscle cells are naturally in a relaxed state.

3 YOUR MUSCLES are packed with LONG, THIN CELLS.

They contract (shorten) when they receive a signal from a nerve cell.

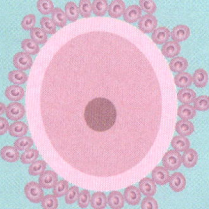

4 FAT CELLS are like BUBBLES FILLED WITH OIL.

They get bigger or smaller depending on how much fat is stored by the body.

5 EPITHELIAL CELLS LINE YOUR nose, mouth, lungs, and digestive system.

They also make up your outer, protective layer: skin.

Cell nucleus (see p16)

21 October
EXTREME DINOSAURS

1
T-REX HAD THE STRONGEST BITE OF ANY LAND ANIMAL that has ever lived.

Its powerful jaw muscles and 60 sharp, serrated teeth could tear flesh and crush bone.

2
THERIZINOSAURUS had the LONGEST CLAWS of any dinosaur.

In spite of its 70-cm (2.3-ft) long, sickle-shaped claws, it was probably a herbivore.

Therizinosaurus may have used its claws for pulling down high tree branches.

3
THE LARGEST DINOSAURS WERE SAUROPODS.

These herbivorous, long-tailed, long-necked beasts could reach 37 m (122 ft) from nose to tail.

Patagotitan mayorum was the longest dinosaur and may have been the largest land animal of all time.

5
PENTACERATOPS had the LARGEST HEAD.

Its skull, including a huge, bony head frill, could grow to 3.2 m (10 ft) long.

4
ARGENTINOSAURUS WEIGHED 100 tonnes (110 tons) – AS MUCH AS 16 ELEPHANTS.

22

DECOMPOSERS

This microscope image shows soil bacteria magnified 1,500 times.

1

DECOMPOSERS are organisms that FEED ON DEAD MATTER AND FAECES (poo).

The most important decomposers are bacteria, which are microscopic, and fungi.

Mushrooms are just the fruiting parts of fungi that live in soil or in rotting wood.

2

Decomposers are NATURE'S RECYCLERS.

They break down large organic molecules (such as proteins) into simple nutrients that plants can absorb and use again (such as nitrates).

Soil mites are related to spiders but are much smaller. They eat all sorts of dead matter.

Earthworms eat dead leaves. Their droppings help plants grow.

3

SOIL is FULL of DECOMPOSERS.

A single teaspoon of soil can contain more than 1 billion bacteria.

Springtails are insect-like animals that live in soil. They break down dead matter into smaller pieces.

4

Decomposers make COMPOST.

Gardeners put dead plants and kitchen waste into compost bins so that decomposers can recycle it into plant food.

5

FRIDGES SLOW DOWN DECOMPOSITION.

Low temperatures delay the growth of bacteria and fungi in food.

1 ELECTRONIC GOODS, such as PHONES, TVs, and speakers, all CONTAIN MAGNETS.

So do many household appliances, and even medical equipment such as MRI machines (see p150).

2 NEODYMIUM MAGNETS are the STRONGEST magnets you can buy.

They are used in many electronic and medical devices because they keep their magnetism for much longer.

3 A FRIDGE MAGNET is much stronger than EARTH'S MAGNETIC FIELD (see p165)!

Maglev trains use magnets instead of wheels, which reduces friction.

4 MAGLEV TRAINS CAN LEVITATE using magnetic forces.

Under the train, magnetic attraction and repulsion are also used to propel the train forward at speeds of up to 603 km/h (375 mph).

5 Large SCRAPYARD MAGNETS are strong enough to LIFT OBJECTS weighing tonnes, such as cars.

Smaller ones can separate out metals from other materials for recycling.

23
October
MAGNETS IN ACTION

1 WAVES are VIBRATIONS that TRANSFER ENERGY as they travel.

2 LIGHT and other waves on the electromagnetic spectrum (see p204) travel in TRANSVERSE WAVES.

Transverse waves move up and down in an S-shaped pattern.

24
October
TYPES OF WAVE

4 WATER WAVES ARE MADE OF TWO TYPES OF WAVES.

Surface waves are transverse, but under the water they are longitudinal.

5 EARTHQUAKES send out shockwaves through Earth called SEISMIC WAVES (see p58).

Particles in transverse waves move up and down at right angles to the wave's direction.

Amplitude measures the strength of a wave.

Wavelength

3 SOUND travels in LONGITUDINAL WAVES

These waves squeeze and stretch the air molecules like a spring (see p231).

A wave transfers energy, not particles.

Resting position

1

THE SUN IS A STAR.
This giant ball of glowing gas sits at the centre of our Solar System, and holds all the objects – from tiny asteroids to vast planets – in its orbit.

2

IT IS 1.4 million km (870,000 miles) WIDE.
Even though it's 100 times wider than Earth, the Sun is just an average size compared to other stars.

3

The Sun's CORONA (outer atmosphere) is THOUSANDS OF TIMES HOTTER than its SURFACE.
But scientists aren't sure why! It can reach temperatures of up to 2 million°C (3.5 million°F).

26
October
THE SUN

Energy diffuses outwards through the radiative zone.

Energy rises to the Sun's surface, then sinks back down in the convective zone.

In the core, hydrogen is converted to helium in a reaction called nuclear fusion. These reactions power the Sun and make it shine.

The Sun has three layers of atmosphere. The photosphere is the innermost layer, then the chromosphere, and then the corona.

Loops of electrically charged gas, called prominences, extend out from the Sun's surface.

4

The Sun is LESS THAN HALFWAY THROUGH ITS LIFE.
It is about 4.5 billion years old and will last about 5 billion more years before it transforms into a white dwarf (see p99).

5

THE SUN ORBITS THE CENTRE OF THE MILKY WAY GALAXY (see p200).
Along with the rest of the Solar System, it travels at about 720,000 km/h (450,000 mph) and takes around 230 million years to complete one orbit.

27
October
EARLY SURGERY

1 People have been PERFORMING SURGERY SINCE THE STONE AGE.

There is evidence of a leg amputation from 31,000 years ago, as well as holes being drilled in skulls from 8,000 years ago (see p22).

2 Early surgery was performed with knives MADE FROM FLINT (stone) AND OBSIDIAN (volcanic glass).

In the 18th century, patients risked bleeding to death and infection.

3 THE CHAINSAW WASN'T INVENTED FOR CUTTING DOWN TREES.

A small hand-powered chainsaw was first used in the late 18th century to help with childbirth.

4 Patients used to be AWAKE for SURGERY! Anaesthetics weren't invented until the 1840s.

5 SURGEONS DIDN'T WASH THEIR HANDS OR TOOLS BEFORE OPERATING until the 19th century.

Before then, no one knew that microorganisms such as bacteria caused infection.

The patient had to be held down, as they were still awake.

---

1 TORNADOES ARE VIOLENTLY ROTATING COLUMNS OF AIR that form beneath storm clouds.

2 WINDS IN A TORNADO CAN REACH SPEEDS OF 480 km/h (300 mph).

That's strong enough to uproot trees and pick up heavy objects like cars.

3 About 1,000 TORNADOES HAPPEN EVERY YEAR IN THE US, most of them in a place called Tornado Alley.

28
October
TORNADOES

4 A TORNADO that occurs OVER WATER is called A WATERSPOUT.

5 Tornadoes are MEASURED USING THE FUJITA SCALE (F-Scale), and range from F0 to F5.

F5 is the strongest and most dangerous, and can demolish large buildings or suck a riverbed dry.

A sudden change in wind speed or direction (or both) creates a rotating tube of air.

During the day, the Sun heats the ground. Rising draughts of warm air tilt the tube up and a cloud starts to form.

As the cloud grows, a storm begins and starts to spin. Cold air reaches down to the ground as a funnel.

Heavy rain   Downdraught

1 NOT ALL ATOMS (see p53) are STABLE.

Some atoms break down and release energy, called radiation, in the form of electrons. This is known as radioactivity.

An electron is lost from the nucleus as the atom "decays".

2

There are THREE TYPES of radiation: ALPHA, BETA, and GAMMA.

Alpha radiation is the weakest and gamma radiation is the strongest. Gamma rays are high-energy electromagnetic waves (see p204).

Alpha particles cannot pass through a sheet of paper.

Beta particles can be stopped by a sheet of aluminium.

Gamma particles can be stopped only by thick materials, like many centimetres of lead or concrete.

28
October
RADIOACTIVITY

3 French scientist Henri Becquerel DISCOVERED RADIATION in 1896.

He put a piece of uranium on photographic film covered with lightproof paper and discovered the film had an image of uranium on it. This meant something must have come from the uranium.

4

MARIE AND PIERRE CURIE won the NOBEL PRIZE FOR PHYSICS in 1903 for their work on radiation.

The French scientists discovered a new element 400 times more radioactive than uranium, and named it polonium.

5 RADIOACTIVE DATING helped scientists work out the AGE OF THE SOLAR SYSTEM.

They measured the radioactive decay (breakdown) of material in meteorites from the formation of the Solar System.

Atoms in the meteorite decay at a steady rate.

Over time, number of decayed atoms increases

Ratio of different materials reveals age

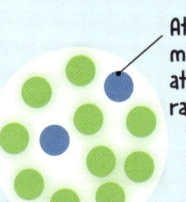

3 BILLION YEARS AGO

2 BILLION YEARS AGO

1 BILLION YEARS AGO

29
October
STEEL

1 Steel is an ALLOY – a mixture made from TWO OR MORE ELEMENTS (see p152).

At least one of the elements is a metal.

2 MILLIONS OF TONNES of steel are RECYCLED each year.

This makes it one of the most recycled materials in the world.

3 To make steel, IRON ORE is melted, its IMPURITIES are extracted, and it is then MIXED with a little CARBON.

4 Adding some CHROMIUM to iron makes STAINLESS STEEL.

Its durability and heat resistance make it perfect for kitchen equipment.

5 Steel can be 1,000 TIMES STRONGER than iron.

Alloys are generally harder and more resistant to breaking than pure metals.

Steel is used in construction because of its strength.

30
October
ULTRAVIOLET LIGHT

1 ULTRAVIOLET (UV) LIGHT is a type of RADIATION.

It is an invisible wavelength of light on the electromagnetic spectrum (see p204).

2 The SUN EMITS UV light.

Small amounts help us produce vitamin D, but it can burn our skin.

3 UV light is INVISIBLE TO HUMANS, but bees, butterflies, and some other insects can see it.

4 UV light can be used to STERILIZE medical equipment as it KILLS OFF harmful bacteria and viruses.

5 Teeth, urine, some minerals, and even platypuses all GLOW UNDER UV LIGHT.

1 OCEANS cover more than 70 per cent of Earth's surface and CONTAIN MORE LIFE THAN ANY OTHER NATURAL HABITAT.

31 October
OCEAN DEPTHS

2 Temperature, pressure, light, and oxygen levels CHANGE DRAMATICALLY in the ocean WITH DEPTH.

These factors determine the creatures that live at each depth.

3 Scientists HAVE DIVIDED THE OCEAN into FIVE DIFFERENT HABITAT ZONES BY DEPTH.

Sperm whales can take a breath of air at the surface and dive down into the midnight zone to catch food for up to two hours!

SUNLIGHT ZONE
0–200 M (0–660 FT)

TWILIGHT ZONE
200–1,000 M
(660–3,300 FT)

MIDNIGHT ZONE
1,000–4,000 M
(3,300–13,100 FT)

The only light in the midnight zone comes from animals who produce it themselves (see p40).

ABYSSAL ZONE
4,000–6,000 M
(13,100–19,700 FT)

4 THE DEEPEST PART OF THE OCEAN IS AT THE BOTTOM OF THE MARIANA TRENCH in the Pacific Ocean, which is 10,994 m (36,070 ft) down.

It's deeper than Mt Everest is tall!

5 Creatures in the hadal zone have SPECIAL ADAPTATIONS TO SURVIVE.

Chemicals in their cells stop them being crushed by the weight of the water above.

Most food in the deep-sea trenches has drifted down from the levels above.

HADAL ZONE
MORE THAN 6,000 M
(19,700 FT)

1
November
SPEED

If you know how far the cyclist travelled and how long it took, you can calculate how fast they were cycling.

DISTANCE

SPEED | TIME

Speed = Distance ÷ Time

1 SPEED is a measure of HOW FAST an OBJECT IS MOVING, while velocity is an object's speed in a specific direction.

2 SPEED can be MEASURED with a STOPWATCH or A SPEEDOMETER.

3 ACCELERATION IS AN OBJECT'S CHANGE IN SPEED.
It can be either positive (speeding up) or negative (slowing down).

4 An object's average SPEED can be WORKED OUT BY DIVIDING the DISTANCE TRAVELLED by the TIME it took TO TRAVEL THE DISTANCE.

5 The FASTEST THING in the Universe is LIGHT - reaching speeds of 300,000 km/s (186,000 miles/s).

---

1 The visible part of your nail is MADE OF DEAD CELLS hardened by a SPECIAL PROTEIN.
Underneath this is a living layer, made of different skin cells, called the nail bed.

2 IF YOU DIDN'T CUT (OR CHEW) YOUR NAILS your whole life, each one would be about 3.4 m (11 ft) LONG!

The cuticle protects the nail as it grows.

3 FINGERNAILS GROW FOUR TIMES FASTER THAN TOENAILS.
And fingernails grow faster on the hand you use most.

The nail plate is hardened by a special protein called keratin.

4 ONLY HUMANS, MONKEYS, AND APES HAVE FINGERNAILS.
Other animals have claws.

5 THE MEDICAL NAME FOR the habit of NAIL BITING is ONYCHOPHAGIA.

The nail bed has blood vessels to feed the growing nail.

The root, where growth occurs, is hidden under the cuticle.

2
November
YOUR NAILS

2

TARDIGRADES are microscopic creatures that survive extreme conditions by **SHRINKING.**

They squeeze out the water from their body and retract their head and legs. When conditions improve, the tardigrades go back to normal.

Tardigrades are just 0.5 mm (0.02 in) long.

1

The **WOOD FROG** survives the winter by **FREEZING OVER.**

Freezing would be deadly for most creatures, but the wood frog produces a substance that stops the water inside its cells freezing.

The frog's heart and breathing stop when it freezes, then restart when it thaws.

3

SAHARA DESERT ANTS can withstand temperatures of up to 60°C (140°F).

Long legs keep its body far off the hot ground.

3

November

EXTREME ADAPTATIONS

5

The **OLM** is a salamander that lives in **DARK CAVES.**

It is blind, but has heightened senses of smell and hearing, and can detect electrical signals from its prey.

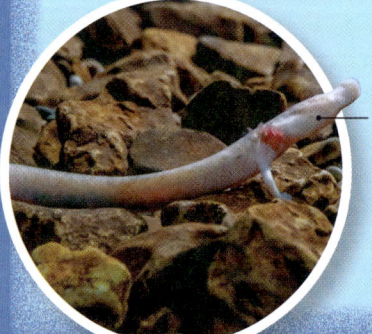

Over time, the olm's skin grows over its eyes, since they are useless in the darkness of the caves.

4

The **THORNY DEVIL** is a lizard that **LIVES IN THE HOT DESERTS** of Australia.

It uses its skin to draw up water from its surroundings and pass it along tiny channels to its mouth.

The water the thorny devil absorbs travels along channels in its scales.

224

4
November
WEATHERING AND EROSION

1 WEATHERING AND EROSION are natural processes that WEAR AWAY ROCK.

Erosion makes hills and mountains slowly crumble away.

2 WEATHERING is the BREAKDOWN OF ROCK INTO SMALL BITS, such as sand particles.

Rain, heating and cooling, ice, and plants cause weathering.

3 EROSION CARRIES AWAY the rock particles made by weathering.

Rivers, glaciers, waves, and wind all cause erosion.

4 Erosion by rivers creates V-SHAPED VALLEYS.

But erosion by glaciers creates U-shaped valleys.

Rivers carry sand and pebbles downstream.

5 EROSION by the Colorado River CREATED THE USA'S GRAND CANYON.

It took 5 million years to form.

5
November
MIXTURES

1 A MIXTURE is where different materials are MIXED together, but can be EASILY SEPARATED.

Sieving a mixture can separate small particles from larger ones.

2 The substances in a mixture DO NOT REACT TOGETHER OR CHANGE CHEMICALLY.

When two substances bond together, they are called a compound.

3 Mixtures can be SEPARATED by several methods, such as SIEVING, FILTERING, OR EVAPORATING.

Filtering can separate out small particles from a liquid.

4 A SOLUTION is a mixture where a solid or gas COMPLETELY MIXES (DISSOLVES) in a liquid.

Salt dissolves in water to form a solution.

Heating a solution can cause the liquid in it to change to a gas, leaving the solids behind. This is called evaporation.

5 ICE CREAM is a mixture!

It is a mixture of fat, water, sugar, ice, and air bubbles.

1 PUBERTY is the process the human body goes through IN ORDER TO BECOME SEXUALLY MATURE.

It usually begins roughly between the ages of 8 to 13 in girls, and 9 to 14 in boys.

2 TWO MAIN HORMONES (see p257) CONTROL PUBERTY.

Girls begin releasing oestrogen and boys begin releasing testosterone.

3 As well as physical changes such as developing breasts and pubic hair, GIRLS GET THEIR FIRST MONTHLY PERIODS.

4 IN BOYS, facial, body, and pubic HAIR GROWS; the testes and penis enlarge; and the CHEST and SHOULDERS BROADEN.

5 BOYS' VOICES GET DEEPER during puberty.

Testosterone makes the vocal cords in the throat grow longer and thicker, causing them to vibrate more slowly and produce a lower pitch.

6
November
PUBERTY

1 SCIENTISTS AREN'T exactly SURE WHAT CAUSES AGEING.

Damage to cells and DNA accumulates over time, but it is unclear why older bodies are less able to repair this.

7
November
AGEING

2 THE WORLD IS AGEING!

Latest trends show that the amount of people above the age of 65 years is growing more rapidly than the population below that age.

3 Signs of ageing in humans include GREY or thinning HAIR, WRINKLY SKIN, MUSCLE LOSS, and LESS sensory ability.

4 Research is underway to SLOW or REVERSE AGEING.

One idea is looking at reprogramming types of white blood cells in the body.

5 The naked mole rat can repair its cells so well and quickly that it NEVER SHOWS SIGNS OF AGEING.

1 The changes that a living thing goes through are CALLED ITS LIFE CYCLE.

Some living things (including humans) grow into bigger, stronger versions of their younger selves.

6. Once the transformation is complete, the adult butterfly emerges. It will be ready to fly within two hours.

2 Other organisms go through COMPLETE METAMORPHOSIS.

This is where their adult form is totally different to their younger ones. A butterfly undergoes complete metamorphosis.

5. When the caterpillar has eaten enough, it builds a hard case, called a pupa or chrysalis, where complete metamorphosis takes place.

1. The cycle starts with an egg, which contains just enough food for the development of a small larva.

8
November
LIFE CYCLES

2. Within a week, the larva, called a caterpillar in moths and butterflies, emerges and eats its egg case.

5

Inside a chrysalis, the caterpillar's body is BROKEN DOWN into A LIQUID.

New body parts grow from this goo until, after a few weeks, a butterfly emerges.

3. The caterpillar is then on a mission to eat as much as it can. As it grows, it will shed its skin at least five times.

3

ADULT MAYFLIES have a VERY SHORT LIFESPAN.

Mayflies can spend years as nymphs, but adults live for just one day.

4. The caterpillar will eat several times its own body weight as it grows, and develops.

4

Periodic cicadas SPEND UP TO 17 YEARS UNDERGROUND as wingless nymphs.

The flying adult cicadas emerge in huge swarms in the spring.

9
November
GRASSLANDS

1 Grasslands occur in places that are TOO DRY for FORESTS, but NOT DRY ENOUGH for DESERTS.

They usually lie in the hearts of large continents, far from the ocean.

2 There are two types of grassland: TROPICAL and TEMPERATE.

Tropical grassland is hot all year round, while temperate grassland has cold winters and hot summers.

The acacia has thorns the size of bananas, and is one of the few trees tough enough to survive here.

Large ears help African elephants lose excess heat.

Cheetahs can reach top speed in three seconds – perfect for chasing speedy antelope.

3 Grasslands make up about 30 PER CENT of the WORLD'S LAND.

They exist on every continent except Antarctica.

Crocodiles attack animals that come to drink the water.

4 Grass attracts GRAZERS who, in turn, attract PREDATORS!

Many grazers, such as wildebeest and zebra, live in large herds for protection against predators.

Meercats stand on their back legs to keep watch for predators.

Pythons kill their prey by squeezing them to death.

5 In such an EXPOSED HABITAT, many grassland animals seek SAFETY UNDERGROUND.

Termites build mounds with networks of tunnels and chambers that extend underground.

Mole rats burrow to find food, living off the roots, bulbs, and tubers of plants.

An aardvark uses its long tongue to lap up termites.

10
November
HAIL

The ice pellets grow larger as more layers of ice are added.

Air currents carry the frozen water droplets up and down.

Rising air

1 Hailstones are SOLID PELLETS of ICE that form in COLD STORM CLOUDS.

2 Hail forms when FROZEN WATER DROPLETS in the air are DRAGGED up and down inside a storm cloud.

Ice falls as hailstones

3 As the hailstones MOVE through the cloud, they build up LAYERS OF ICE.
They eventually become so heavy that they fall to the ground.

4 HUGE hailstones can form when smaller hailstones SMASH and STICK TOGETHER.

Large hailstones form when the air is very humid and the updrafts in a cloud are strong.

5 The LARGEST known hailstone was 20 cm (8 in) across. It fell in South Dakota, USA, on 23 July 2010.

11
November
MICROWAVES

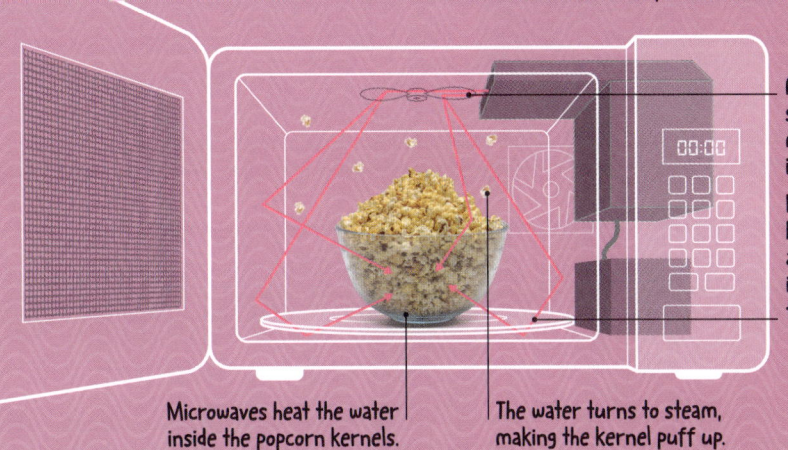

Microwaves heat the water inside the popcorn kernels.

The water turns to steam, making the kernel puff up.

A fan scatters the microwaves into the oven.

Microwaves bounce around inside the oven.

1 Microwaves are a type of INVISIBLE LIGHT.
They are part of the electromagnetic spectrum (see p204).

2 Microwaves can TRAVEL THROUGH WALLS.
This makes them useful in communication, such as Wi-Fi.

3 Microwave OVENS use MICROWAVES.
The waves make water molecules in food vibrate, which heats the food.

4 The microwave oven was INVENTED by ACCIDENT.
A scientist working with microwaves discovered the chocolate bar in his pocket had completely melted.

5 Microwaves are USED IN SPACE.
Scientists use them to send messages to spacecraft millions of kilometres away.

229

12

November
CONSERVATION

1 CONSERVATION is the PROTECTION OF THE NATURAL WORLD, including water, soil, minerals, wildlife, and forests.

2 As the human population has increased, MORE LAND HAS BEEN TAKEN OVER for buildings, roads, and farming. This puts WILD ANIMALS AND PLANTS AT RISK.

3 More than 40 PER CENT of the Amazon rainforest is now UNDER CONSERVATION MANAGEMENT OR PROTECTED BY LAW.

Park rangers and guides help visitors to enjoy national parks without harming the wildlife.

4 There are 100,000 WILDLIFE RESERVES AND NATIONAL PARKS worldwide that offer PROTECTION to animals, plants, and their habitats.

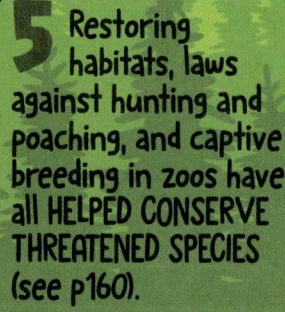

5 Restoring habitats, laws against hunting and poaching, and captive breeding in zoos have all HELPED CONSERVE THREATENED SPECIES (see p160).

1

WEATHER is the CONDITIONS IN THE ATMOSPHERE AROUND US at any one time – a mix of temperature, pressure, and the clouds and wind.

Scientific instruments are tied to weather balloons to record atmospheric conditions.

13

November
WEATHER

2 METEOROLOGISTS study, measure, and try to PREDICT THE WEATHER.

They use equipment such as weather balloons, satellites, and aircraft to collect data.

3 About 67,000 WEATHER BALLOONS are launched in the USA EACH YEAR.

They help us predict the weather by measuring the conditions in the stratosphere (see p26).

4 Three-day weather forecasts are about 97 PER CENT ACCURATE.

Even with modern technology, predicting the weather further in advance is still difficult.

5 PINE CONES RESPOND to the weather.

They open up on dry days so their seeds can be carried away on the breeze, and remain closed when the air is humid.

Sound waves travel
in all directions.

14
November
SOUND

1

All SOUNDS are created by an OBJECT VIBRATING, such as a guitar string.

The vibrations spread into the air, causing invisible waves (see p217) that ripple outwards in every direction until they hit our ears.

2 Sound waves STRETCH and SQUEEZE AIR as they travel through it.

First they squeeze the air molecules together, then they stretch them apart. The more energy the wave has, the more it stretches and squeezes air, and the louder the sound is.

Air particles are spread apart
and under low pressure.

Air particles are compressed
and under high pressure.

3 The NUMBER OF WAVES PER SECOND is known as the FREQUENCY of a sound wave.

High-frequency sounds are heard as a high-pitched sound.

The distance between two
waves is called the wavelength.

HIGH FREQUENCY

Long wavelengths produce
waves with a low frequency.

LOW FREQUENCY

4 Sound waves can TRAVEL THROUGH SOLIDS, LIQUIDS, and GASES, but not through a vacuum.

Outer space is a vacuum (see p145), so it's totally silent.

5 Sound waves travel at about 1,200 km/h (750 mph) in air, but LIGHT IS ABOUT A MILLION TIMES FASTER.

That's why you see lightning a few seconds before you hear the crash of thunder.

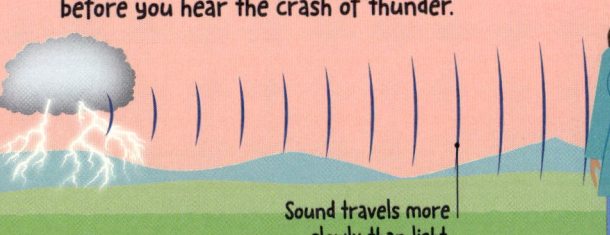

Sound travels more
slowly than light.

1 A fungus is a SIMPLE ORGANISM that is NEITHER PLANT NOR ANIMAL.

Fungi get their nutrients by absorbing them from the environment around them. Mushrooms, moulds, and yeasts are all examples of fungi.

A mushroom or toadstool is known as a "fruiting body".

Spores are released from gills.

2 A fungus usually REPRODUCES by creating seedlike SPORES.

Wind, water, or insects spread the spores, and they grow if they land in the right environment.

3 Fungi COLONIZED LAND 1.3 billion years ago.

They can be found in almost all ecosystems on Earth – even Antarctica.

When they land, spores grow into strands called hyphae.

The mycelium push the fruiting body above the ground.

Hyphae meet and fuse into a network called mycelium.

4 Some fungi are LIFESAVING! The antibiotic PENICILLIN is made from a FUNGUS.

5 More than 125 SPECIES of fungi GLOW IN THE DARK.

Scientists think this bioluminescence (see p40) might be to attract insects, who in turn spread the fungus's spores.

1 There are more than 130 MILLION pieces of SPACE JUNK orbiting Earth.

Altogether they weigh over 9,000 tonnes (9,920 tons).

2 Old SATELLITES, discarded ROCKET PARTS, and smaller fragments from COLLISIONS make up most of the debris.

3 Most space junk is smaller than 10 cm (4 in) wide, but there are more than 36,500 LARGER OBJECTS.

Gravity pulls space junk into orbit around Earth.

16
November
SPACE JUNK

The smallest items are flecks of paint from spacecraft, dislodged during take-off or impacts with other space junk.

4 Pieces of junk can TRAVEL at SPEEDS of 29,000 km/h (18,000 mph).

Even tiny pieces can damage spacecraft and satellites.

5 Scientists are exploring ways we could CLEAR UP the junk put into space, such as using MAGNETS or even a HUGE NET!

---

1 MUSIC is the ARRANGEMENT of sounds in a way that sounds PLEASING.

2 Musical INSTRUMENTS create sounds by making the AIR PARTICLES VIBRATE.

You hear the music when vibrations travel into your ear (see p231).

3 Different WAVELENGTHS form different musical NOTES.

Short wavelengths (see p217) create high-pitched notes, and long wavelengths are low-pitched.

17
November
MUSIC

Drawing a bow across a violin's strings causes them to vibrate.

Sound waves travel out through holes in the body of the violin and through the air to the listener.

4 Each violin STRING can produce many DIFFERENT NOTES.

Pulling a string tighter so it vibrates faster or pressing a finger on a string to shorten it can create a higher note.

5 Humans have been MAKING MUSIC for TENS OF THOUSANDS OF YEARS.

The oldest musical instrument yet discovered is a 60,000-year-old bone flute from Slovenia.

Vibrations travel into the hollow body through a bridge.

SOUND WAVES

The violin's body vibrates, which produces most of the sound.

233

18 November
SEASONS

1

There are FOUR typical seasons in a year: SPRING, SUMMER, AUTUMN, and WINTER.

Leaves sprout and trees blossom when the hours of sunlight increase in spring.

5

ANIMALS follow the seasons closely. Many HIBERNATE or MIGRATE during winter, and RAISE THEIR YOUNG during spring.

In summer, many flowers bloom and fruits begin to ripen.

2

Some parts of the world have FEWER SEASONS.

The poles experience just summer and winter. Near the equator, the temperature stays warm all year round, but there are wet and dry seasons.

In winter, many trees, called deciduous trees, lose their leaves.

Leaves turn to shades of orange and brown when the hours of sunlight decrease in autumn.

4

In summer, the hours of daylight increase until the SUMMER SOLSTICE.

This is the "longest day" of the year: the day with the most hours of sunlight.

3 EARTH has seasons because it SPINS on a TILTED AXIS, meaning different parts of the planet are ANGLED AWAY FROM THE SUN at different times of the year.

Sunlight

Summer in the southern hemisphere occurs when the North Pole tilts away from the Sun.

It is summer in the northern hemisphere when the South Pole tilts away from the Sun.

Earth's orbit

19
November
GEMS

1 There are more than 5,000 KNOWN MINERALS on Earth, but fewer than 100 are used as GEMSTONES.

2 GEMS are usually WEIGHED by the CARAT, which is equal to one-fifth of a gram (0.007 oz).

3 The QUALITY and value of DIAMONDS and other precious gems is measured by the four Cs: COLOUR, CLARITY, CUT, and CARAT.

4 When CUTTING transparent gemstones, the aim is to maximise their "FIRE" and "BRILLIANCE".

"Fire" is how much a gem fractures light into rainbows, and "brilliance" is the amount of light reflected out of the gem.

5 Over THREE-QUARTERS of all diamonds today are cut into a "ROUND BRILLIANT" shape.

Gems are cut into different shapes.

ROUND BRILLIANT

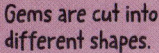

CUSHION

PEAR

STEP

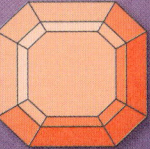
SCISSORS

Emerald-cut gems are rectangular.

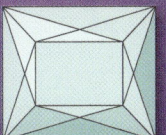
EMERALD

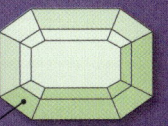

20
November
SILVER

Silver embedded in quartz

1 Silver is a METALLIC ELEMENT (see p152) that can be found in EARTH'S CRUST.

2 When EXPOSED TO AIR, silver will slowly REACT with other substances and TURN BLACK.

3 Pure silver is SOFT. It is often MIXED with small amounts of COPPER to make it HARDER, so it can be used to make items like JEWELLERY.

4 Silver is the best CONDUCTOR of ELECTRICITY out of all the metals, making it useful in ELECTRONIC DEVICES.

5 Silver KILLS BACTERIA, so it can be used in MEDICINE to CLEAN CUTS and TREAT BURNS.

21
November
OPTICAL ILLUSIONS

1 OPTICAL ILLUSIONS are IMAGES DESIGNED TO TRICK US into perceiving something differently from how it really is.

2 When your SENSES TAKE IN INFORMATION, they SEND IT TO YOUR BRAIN.

The brain fills in any gaps to work out what you are seeing. But your brain can get it wrong!

3 Some illusions make IDENTICAL OBJECTS appear to be DIFFERENT SIZES, or make STRAIGHT LINES seem BENT OR SKEWED.

The two red lines are the same length, but do not immediately appear to be.

The vertical parallel lines look like they are tilting away from each other due to the effect of the other lines.

4 Other illusions can make STATIC IMAGES appear to be MOVING, as on the left.

The way shapes, curves, and colours are placed creates this effect.

5 Some FAMOUS ARTISTS, such as MC Escher and Bridget Riley, have USED OPTICAL EFFECTS IN THEIR WORK to challenge the viewer.

22
November
LENSES

1 LENSES are pieces of transparent material with CURVED SURFACES that REFRACT, or change the path, of LIGHT.

2 REFRACTION happens because LIGHT TRAVELS SLOWER through solids than air.

When light enters or exits a solid at an angle, this changes the light's path, so it seems to bend.

Convex lenses make light rays converge (come together).

Focal point

Ray of light

CONVEX LENS

3 CONVEX lenses FOCUS light.

They are used in cameras to capture an image, and in microscopes and telescopes to magnify the view.

4 There is a convex lens IN YOUR EYE (see p33).

It is made of clear proteins, and it focuses light by changing shape.

Concave lenses make light rays diverge (spread out).

CONCAVE LENS

5 ARTIFICIAL LENSES made of plastic can be IMPLANTED in people with vision problems.

Millions of people around the world undergo this procedure each year.

23
November
WORMS

Peculiar-looking Christmas tree worms live on coral reefs.

1 There are more than 40,000 KNOWN SPECIES OF WORMS.

Many are microscopic nematodes that live in the soil or the sea bed.

2 EARTHWORMS are important DECOMPOSERS (see p216) that help to break down and recycle dead matter.

An earthworm can eat its own body weight in soil in one day.

3 There are many species of PARASITIC WORMS.

An estimated 300 different species can infect humans alone.

Earthworms have tiny bristles along their body that help them move along.

4 Many types of worm can GROW A NEW TAIL if the original one is broken off.

Some species can even grow into two new worms if cut in half.

5 The BIGGEST EARTHWORM is Australia's Giant Gippsland Earthworm.

It grows to about 1m (3ft) long and weighs 500g (1lb).

24

November
DARK MATTER AND ENERGY

1 Less than FIVE PER CENT of the Universe is made up of ORDINARY MATTER that we can see.

The rest is made up of mysterious substances known as dark matter and dark energy.

2 DARK MATTER is completely INVISIBLE.

Scientists aren't really sure what it is! But they do know that it does not emit, reflect, or absorb light, and that it takes up space and has mass, like matter.

3 Scientists know dark matter EXISTS because of HOW IT INTERACTS with things we can see.

They use giant detectors buried up to 2 km (1.2 miles) underground to look for dark matter passing through our planet.

4 Galaxies ROTATE FAST, but the matter we can see couldn't possibly EXERT enough GRAVITY to hold them together.

Scientists think there must be some unseen mass – dark matter – interacting with them and holding them together.

5 Scientists think an unknown source of ENERGY is making the Universe EXPAND at an INCREASING SPEED.

This energy, known as dark energy, seems to oppose gravity – it is pushing the Universe apart instead of pulling it together.

1 EXTINCTION is when the LAST REMAINING member of a species DIES and no more are left.

2 Extinction can be caused by new PREDATORS, DISEASE, CLIMATE CHANGE, HABITAT LOSS, or a single CATASTROPHIC EVENT.

3 There have been FIVE MASS EXTINCTIONS in the past 600 million years. Each WIPED OUT large portions of the planet's species.

4 The BIGGEST mass extinction happened about 250 MILLION YEARS AGO. About 90 per cent of all species on Earth were wiped out by volcanic eruptions.

5 One of the first animals that HUMANS made EXTINCT was the GIANT GROUND SLOTH, 11,000 years ago.

The slow-moving giant ground sloth stood 7 m (23 ft) tall and weighed 6.3 tonnes (7 tons).

25
November
EXTINCTION

---

1 Glaciers are SLOW-MOVING RIVERS OF ICE that form where SNOW COMPRESSES and TURNS INTO ICE.

2 Glaciers are the LARGEST RESERVOIR of FRESH water on our planet. They store about 68 per cent of the world's fresh water.

3 The LARGEST GLACIER is Antarctica's Lambert-Fisher Glacier, which is 900 FOOTBALL PITCHES WIDE and 3,000 LONG!

26
November
GLACIERS

4 Glaciers can appear BLUE in colour when they become very DENSE and free of AIR BUBBLES.

5 During the last ice age, glaciers COVERED most of NORTH AMERICA and much of NORTHERN EUROPE. Today, most are found near the poles, but also high up mountains around the world.

Fresh snow adds to the glacier.

The glacier is pulled downhill by gravity.

The glacier erodes rocks and later deposits them in piles called moraines.

The ice cracks, forming crevasses.

Sometimes a small glacier will join a larger one.

The glacier ends at a lake or ocean. Chunks break off as icebergs.

1 HAIR IS MADE OF A HARD PROTEIN CALLED KERATIN.

This is the same stuff your nails are made of (see p223), and can be found in your outer layer of skin.

Inside the cuticle is the cortex, full of many bundles of filaments, which are glued together in a strong, flexible shaft.

Long spirals of keratin lie at the centre of each hair shaft.

Layers of overlapping cuticle cells on the outside protect the hair.

2 HAIR TEXTURE IS DETERMINED BY THE SHAPE OF THE FOLLICLES (the holes that your hair grows out of).

STRAIGHT HAIR | WAVY HAIR | CURLY HAIR

3 RED IS THE RAREST HAIR COLOUR - 2 per cent of all the people on Earth have red hair.

Scotland, UK, has the most redheads - 13 per cent.

27
November
YOUR HAIR

4 HAIR ON YOUR BODY STANDS ON END WHEN YOU ARE COLD OR SCARED.

Each hair is attached to a tiny muscle that pulls at the base to make the hair stand up and trap more air to keep you warm.

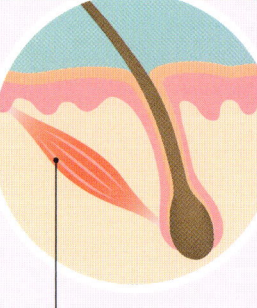

Relaxed muscle

Genetics determines hair colour and texture.

Goose bump

Your head hair grows by about 1cm (0.4 in) a month.

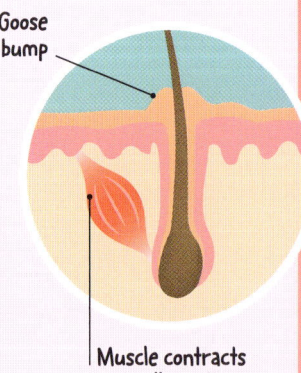

Muscle contracts and pulls on the hair

5 There are about 100,000 HAIRS ON YOUR HEAD, AND 5 MILLION ON YOUR BODY.

Between 50 and 100 head hairs fall out each day, but new hairs are constantly growing to replace them.

Blonde people have the most hairs on their heads - about 130,000.

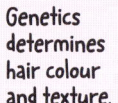

It can take up to six years to grow hair down to your waist.

28 November
SPACE TELESCOPES

2 The JWST can see back TO THE VERY BEGINNING OF THE UNIVERSE! This is because its huge mirror can collect so much light.

3 THE FIRST SPACE TELESCOPE was launched into orbit in 1962. *Ariel 1* measured solar radiation and carried out other experiments.

4 The Hubble Space Telescope was launched in 1990, and HAS MADE OVER 1.6 MILLION OBSERVATIONS.

1 The James Webb Space Telescope (JWST) is THE MOST POWERFUL TELESCOPE EVER SENT INTO SPACE.

5 Space telescopes have discovered EXOPLANETS (see p50), WATER VAPOUR ON OTHER PLANETS, and THE MOONS OF PLUTO.

---

1 WATER COVERS 71 PER CENT OF EARTH'S SURFACE.

The amount of water looks so small but, without it, there would be no life on Earth.

3 JUST THREE PER CENT of Earth's water is FRESH WATER, and most of that is STORED AS ICE in glaciers and ice caps.

2 IF ALL THE WORLD'S WATER WAS FORMED INTO A BALL, it would be just 1,384 km (860 miles) WIDE.

4 ONE-FIFTH OF THE FREE-FLOWING FRESH WATER IS CARRIED BY THE WORLD'S LARGEST RIVER – THE AMAZON.

29 November
WORLD OF WATER

5 AN OAK TREE can release THREE BATHFULS OF WATER INTO THE AIR through its leaves in a day.

The average depth of the world's oceans is about 3.5 km (2.2 miles).

1

There are more than 500 different SPECIES OF PRIMATE.

The group includes monkeys, apes, lemurs, tarsiers, and lorises.

HAINAN GIBBON

TARSIER

2

Just 1.2 per cent of a chimpanzee's DNA is different from a human's.

YOU are a PRIMATE!

Humans and other modern primates descended from a common ancestor millions of years ago. Our closest animal cousins are bonobos and chimpanzees.

3

HOWLER MONKEYS are the world's LOUDEST LAND ANIMAL.

Their hooting calls can be heard 4.8 km (3 miles) away.

30
November
PRIMATES

4

The BIGGEST PRIMATE is the mighty GORILLA.

The smallest is Madame Berthe's mouse lemur, which could fit in the palm of a gorilla's hand.

5

ALMOST ALL MONKEYS HAVE TAILS, while apes do not.

The spider monkey has a long, grasping tail that it uses as a fifth limb to cling onto trees.

An eastern lowland gorilla can reach 1.7 m (5.5 ft) tall.

1
December
PALM TREES

1 PALM TREES are found mostly in TROPICAL REGIONS. They are known for their long, compound leaves known as FRONDS.

2 Palm trees have existed for millennia – even LIVING ALONGSIDE THE DINOSAURS 80 MILLION YEARS AGO.

3 Their leaf structure, flexible trunks, and sturdy root systems make them RESILIENT IN EXTREME WET AND WINDY CONDITIONS.

4 Palm trees have NO BARK! Instead, they have something called pseudobark – dried-out tissue that is left over after their leaf fronds have shed.

5 The COCO DE MER palm tree has the LARGEST AND HEAVIEST SEEDS OF ANY TREE.
These can weigh more than 27 kg (60 lb) and can be up to 50 cm (20 in) wide.

Coco de mer nut

2
December
POND LIFE

1 Although they are only small bodies of water, PONDS CAN SUPPORT MORE BIODIVERSITY THAN SOME RIVERS AND LAKES.

2 POND SKATERS walk on water!
These insects live at the top level of a pond, while fish swim in deeper waters, and scavengers hunt along the bottom.

3 WATER LILIES FLOAT on the surface. They are anchored to the bed by RUBBERY STEMS with tubes that carry air to the roots.

Dragonflies eat tadpoles, mosquitoes, and even fish.

4 Some WATER BEETLES dive taking bubbles of air with them for breathing, and WATER SPIDERS trap air in underwater webs.

5 Many creatures LAY THEIR EGGS in pond VEGETATION.
The smooth newt carefully wraps each of its eggs in a leaf for protection.

1 WATER is a common substance that is ESSENTIAL FOR LIFE.

2 Water is so useful because it can DISSOLVE (see p225) MORE SUBSTANCES THAN ANY OTHER LIQUID.
This means it is able to carry vital minerals, nutrients, and other chemicals.

Solid (ice)

Gas (steam or water vapour)

3 Water can be found in THREE DIFFERENT STATES.
The states are liquid, solid, and gas.

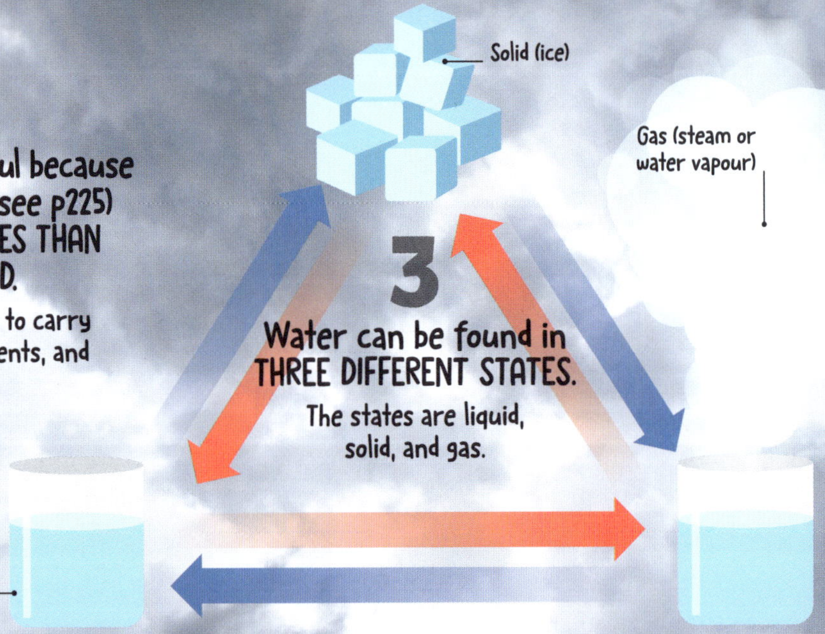

Liquid (water)

3 December WATER

4 Water makes up around HALF of an ADULT HUMAN'S BODY.

5 More than 97 PER CENT of water on Earth is SALTY and found in oceans.

1 CORAL REEFS are the world's LARGEST LIVING STRUCTURES.

2 Corals are actually tiny ANIMALS called POLYPS with hard, stony SKELETONS.

3 Some corals are made of THOUSANDS OF POLYPS that live in a COLONY.

4 About 25 PER CENT of all MARINE ANIMALS live in coral reefs.

5 Corals have GROWTH RINGS like trees!
Scientists can study the layers to learn about the conditions they grew in.

4 December CORAL REEFS

1

Plants make FLOWERS to help them REPRODUCE (see p131).

The bright colours and scents attract pollinating animals. Blue flowers attract insects, and red flowers attract birds.

5

Most flowers are HERMAPHRODITES, which means they are MALE AS WELL AS FEMALE.

However, some flowers are either wholly male or wholly female.

2

Most flowers reward pollinators with FOOD, but some flowers TRICK their visitors.

Bee orchids imitate female bees to attract the male bees that pollinate them.

5
December
FLOWERS

4

A SUNFLOWER is a cluster of MORE THAN 1,000 FLOWERS, called FLORETS.

Florets in the middle of the sunflower head make seeds, while the florets around the rim extend out as petals.

3

BUCKET ORCHIDS use an IRRESISTIBLE SCENT to lure bees into a SLIPPERY TRAP.

As a bee climbs out through an exit hole, pollen is glued on its back.

6
December
YOUR NOSE

1

SMELLS are TINY PARTICLES, called SCENT MOLECULES, which are released into the air by anything that has a smell.

2

We pick up SMELLS when they enter the body through the NOSE.

Special sensor cells inside the nose detect the scent molecules.

3

The nose contains lots of STICKY MUCUS.

The mucus mixes with the scent molecules as they enter the nose. The mucus helps sensor cells in the nose to detect the scents.

The smell of durian is so pungent that the fruit is banned on public transport in Singapore.

Nerves in the olfactory bulb carry information about the smell to the brain.

You have about 12 million sensor cells in your nose.

5

The SENSOR CELLS send the INFORMATION about the scent to the BRAIN.

The brain alerts us if a scent is dangerous, such as the smell of smoke, or if the food we are about to eat smells rotten and might be harmful to us.

Brain

Nose

Scent molecule

Scent molecules can also reach the sensor cells in the nose through the mouth, such as when you are eating food.

4

SMELL is very closely LINKED TO TASTE. Without a sense of smell, many tastes in the food we eat are IMPOSSIBLE TO RECOGNIZE.

1 THE UNIVERSE IS EVERYTHING: all matter, energy, time, planets, stars, galaxies, and the space between them.

2 THE UNIVERSE IS 13.8 BILLION YEARS OLD.
It began with the Big Bang (see p23).

3 WE CAN'T SEE THE WHOLE UNIVERSE.
We can see only the parts that light has had time to travel from since the Big Bang. This is called the "observable Universe".

4 The UNIVERSE IS EXPANDING.
Galaxies are moving further away from each other as it grows.

5 WE DON'T KNOW ITS TRUE SIZE.
All we know is that the observable Universe is at least 93 billion light years across!

---

1 MOMENTUM is a measure of THE "OOMPH" something has AS IT MOVES.
It is based on the relationship between an object's speed and mass.

2 A HEAVIER OBJECT has MORE MOMENTUM.
A car has more momentum than a motorcycle travelling at the same speed.

3 A FASTER OBJECT has MORE MOMENTUM.
If two bicycles are moving along at different speeds, the faster one has more momentum.

8
December
MOMENTUM

Bumper cars have bouncy rubber bumpers, so the cars exchange some momentum in a collision.

4 MOMENTUM CAN BE TRANSFERRED.
When you crash into someone in a bumper car, your momentum is transferred to them, making them and their car jolt.

5 ROCKETS USE MOMENTUM TO GET INTO SPACE.
As exhaust gases shoot out in one direction, the rocket is pushed into the air.

1

PLATYPUSES have a SIXTH SENSE, called ELECTRORECEPTION.

Organs called electroreceptors detect electrical signals given out by prey, so the platypus can hunt even in murky waters.

Electroreceptors in the bill detect the signals.

When an animal moves, its muscles emit weak electrical signals.

2

Each arm is lined with hundreds of suckers.

OCTOPUSES TASTE with their ARMS.

Each sucker on their arms is packed with sensory cells that detect taste and touch.

When an octopus sticks its arm into small spaces looking for prey, it can taste if the food is toxic or safe to eat.

3

FIRE CHASER BEETLES can SENSE FIRE from 130 km (80 miles) away.

They search for fires using special organs that can detect infrared radiation (see p256).

Fire chaser beetles lay their eggs in freshly burnt wood.

9
December
ANIMAL SENSES

4

Some BIRDS have a BUILT-IN COMPASS that helps them "SEE" EARTH'S MAGNETIC FIELD.

A molecule in their eyes can sense the magnetic field, helping them find their way during migration.

5

Katydids have TINY EARS on their FRONT LEGS.

These ears are less than a millimetre long and are some of the smallest ears in the animal kingdom.

The thin membrane vibrates like a human's eardrum (see p178).

1

LIGHT is a WAY OF TRANSFERRING ENERGY. It is made up of particles called PHOTONS, which travel like WAVES (see p217). We see things because light REFLECTS (bounces) off objects and into our eyes.

2 When light is shone into a triangular block of glass called a PRISM, it can be SPLIT INTO DIFFERENT COLOURS (see p194) because a prism bends light.

Each wavelength of light is refracted by a different amount, so they come out of the prism at different angles and can be seen as their different colours.

The beam of light refracts (bends) when it enters the prism.

3

Light travels at DIFFERENT SPEEDS in DIFFERENT MATERIALS.

It travels quickly through air, and slows down when it enters a liquid or solid, so it refracts (bends).

10
December
LIGHT

Beam of light

Prism

Violet is refracted the most and red is refracted the least.

4

The phrase "IN A JIFFY" can refer to an EXACT MEASUREMENT.

In physics, a jiffy is the time it takes for light to travel one femtometre – a millionth of a millionth of a millimetre (3.9370×10^{-14} in) – in a vacuum.

5 Looking at BRIGHT LIGHTS makes some people

SNEEZE!

This is called the photic sneeze reflex. It affects about one in three people.

1

There are more than 33,000 species (unique types) of FISH IN THE WORLD.

The majority are cold-blooded and spend their whole lives in water, apart from mudskippers.

Mudskippers spend most of their life on land. To keep their skin moist, they roll in puddles.

Its special, muscular fins allow it to swim, walk, jump, and climb. In fact, it moves faster on land than in water.

2

Instead of lungs, FISH USE GILLS TO BREATHE.

Gills filter oxygen out of water in a similar way to how the lungs of land animals extract oxygen from air.

Water enters mouth

Feathery gills absorb oxygen from the water, and release carbon dioxide into the water.

Water passes out of gills

11
December
FISH

3

SHARKS AND RAYS DON'T HAVE ANY BONES.

Instead, their skeletons are made of cartilage – the same soft, bendy stuff that your ears and nose are made of.

Cartilage is lighter and more flexible than bone.

4

Most fish swim BY WIGGLING THEIR BODIES FROM SIDE TO SIDE.

Resistance of the water to the sideways movement of the tail and body pushes the fish forwards.

Movement of tail

Forward thrust

Resistance of water

Paired fins moved to steer

Resistance of water

Movement of tail

Forward thrust

FORWARD MOVEMENT

5

Fish use their dorsal fin for balance and steering.

The black marlin is THE FASTEST FISH IN THE WORLD.

It is capable of accelerating to an incredible speed of 129 km/h (80 mph).

12
December
PANDEMICS

1 A pandemic is an outbreak of an infectious disease that SPREADS VERY QUICKLY AND ACROSS MULTIPLE COUNTRIES, often globally.

2 EPIDEMICS are when a disease spreads faster or more widely than usual, but IN A SMALLER AREA.

3 In the 14th century, Europe was a hit by a pandemic of BUBONIC PLAGUE, known as the BLACK DEATH.

It is thought to have killed up to 50 per cent of Europe's population.

4 More people living in cities, global travel, and humans living closer to animals MAKE IT EASIER FOR DISEASES TO SPREAD more quickly today than in the past.

5 Many pandemics begin when a DISEASE JUMPS FROM HUMANS TO ANIMALS.

COVID-19, which was declared a pandemic on 11th March 2020, is thought to have originated in bats.

13
December
PIMPLES

1 WHEN TINY OPENINGS in your skin CALLED PORES BECOME CLOGGED with dead skin cells, pimples can form.

2 PIMPLE PUS is made up of DEAD SKIN CELLS, WHITE BLOOD CELLS, BACTERIA, and a sticky substance called SEBUM.

3 WHITEHEADS ARE PIMPLES FILLED WITH PUS that build up under the skin's surface LIKE A VOLCANO.

The pimple looks red and swells as blood rushes to the area.

Pus

White blood cell

4 PIMPLES SWELL UP because of your body's IMMUNE SYSTEM (see p114).

Bacteria multiply and feed on the excess sebum, the area swells as blood rushes to protect it, and your white blood cells attack the bacteria.

Pimples occur in the top layer of skin, called the epidermis (see p57).

Hair growing in pore

Sebaceous gland makes sebum

5 BLACKHEADS form when a pimple is OPEN TO THE AIR.

The air reacts with the sebum and turns it dark in colour.

251

14

December
PLANTS

1 PLANTS MAKE their own FOOD.
Green parts of plants use sunlight to make food. This process is called photosynthesis (see p136).

2 There are more than 391,000 PLANT SPECIES.

3 Inside FRUIT are SEEDS containing everything needed to grow a new plant.

The flower of a tomato plant becomes a fruit after it has been fertilized (see p131).

4 The STEM supports the plant.

The stem carries nutrients and water to the rest of the plant.

5 ROOTS anchor a plant in the ground.
Water and nutrients are absorbed through the roots (see p103).

15

December
SILICON

Pure silicon is hard and brittle.

1 SILICON is a dark-grey element (see p152).
It is so reactive that pure silicon can't be found in nature. Instead, it forms minerals (see p180).

2 Silicon is the SECOND MOST ABUNDANT element found in EARTH'S CRUST, after oxygen.

3 Silicon can REACT WITH OTHER SUBSTANCES to make SILICONE, a heat-resistant and non-stick material often used in cooking equipment.

Silicone cupcake case

4 COMPUTER PARTS and MICROCHIPS (see p93) use silicon.
It is a semiconductor. This means it has specific electrical properties that make it useful in technology.

Silicon wafers are used in microchips.

5 Some scientists think the Universe could be home to SILICON-BASED LIFE FORMS!
Life on Earth is carbon-based, but other environments in the Universe could favour silicon-based life.

16
December
HIBERNATION

Brown bears build up fat stores over summer to survive hibernation.

1 HIBERNATION is a way for animals to CONSERVE THEIR ENERGY over prolonged periods.

2 A hibernating animal's HEART RATE and BREATHING SLOWS, and its BODY TEMPERATURE DROPS.

3 Animals in COLD PLACES hibernate to SURVIVE HARSH WINTERS, when there is little food available.
Some animals in hot countries undergo a type of dormancy called aestivation, which enables them to survive heat and droughts.

4 Most hibernating animals have periods where they WAKE UP and MOVE AROUND a bit.
Female bears may even wake up from hibernation to give birth!

5 In 2007, an Australian eastern pygmy possum experienced the LONGEST HIBERNATION – MORE THAN A YEAR!

---

1 INVERTEBRATES are animals WITHOUT A BACKBONE or BONY SKELETON.

2 97 PER CENT of all animals are invertebrates, making them the LARGEST GROUP IN THE ANIMAL KINGDOM.
Scientists have categorized around 1.3 million species so far.

3 ARTHROPODS, MOLLUSCS, WORMS, ECHINODERMS, and CNIDARIANS are the main groups of invertebrates.

17
December
INVERTEBRATES

4 Many invertebrates have a HARD EXOSKELETON – an outer skeleton that works like a SUIT OF ARMOUR.
But some, such as squid, have soft bodies, so rely on water for support.

5 CORALS and ANEMONES are invertebrates that FILTER THEIR FOOD out of the water.

Crabs are part of a group of invertebrates called crustaceans.

There are more than 7,000 species of crab.

Crustaceans have a hard exoskeleton.

THE SUN

1 The Solar System formed 4.6 BILLION YEARS AGO from a CLOUD OF DUST AND GAS.

About 99.9 per cent of this material became the Sun. The remaining 0.1 per cent formed the rest of the Solar System.

2 The Sun is so MASSIVE that its gravity (see p75) HOLDS countless SPACE OBJECTS in orbit around it.

They include planets, more than 200 moons, five dwarf planets, and millions of other small objects such as asteroids, and comets.

MERCURY

VENUS

The four inner planets are rocky worlds.

EARTH

18
December
SOLAR SYSTEM

MARS

3 The Solar System includes EIGHT PLANETS: Mercury, Venus, Earth, Mars, Jupiter, Saturn, Uranus, and Neptune.

The four outer planets are gas and ice giants.

JUPITER

All eight planets orbit the Sun in an anticlockwise direction on an elliptical (oval) path.

The Asteroid Belt contains millions of rocky and icy asteroids.

4 The Solar System is 30 trillion km (18.6 trillion miles) wide – that's 200,0000 TIMES the distance BETWEEN EARTH AND THE SUN.

SATURN

URANUS

5 At the edge of the Solar System lies a vast CLOUD of ICY and ROCKY OBJECTS called the OORT CLOUD.

These objects are remnants from the creation of the Solar System, and form the boundary of the Sun's gravitational influence.

NEPTUNE

Pluto is a dwarf planet.

PLUTO

The Kuiper Belt is a disc of gas and dust.

19
December
POO

1 An average human poo contains 10,000 MILLION BACTERIA – half of them alive, and the rest dead.

Most are useful bacteria that live in your gut and help the body digest food.

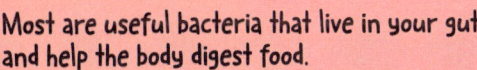

2 You will make about 6 TONNES (6.6 TONS) OF POO in your lifetime – the same weight as a MALE AFRICAN ELEPHANT.

3 It doesn't matter what you eat, HEALTHY POO is always BROWN!

This is because poo contains greenish-brown bile, and reddish-yellow bilirubin, both from your liver. Mixed in the colon, they turn poo brown.

4 But... what you eat DOES affect how poo smells.

When gut bacteria digest food, they emit smelly gases. Highly processed foods take bacteria longer to digest, creating more – and smellier – gas!

5 Thousands of HOUSEHOLDS in Norway are HEATED by POO POWER!

A sewage plant in Oslo sucks heat from sewers to warm up a network of hot water pipes across the city.

20
December
PLANKTON

1 Plankton are a collection of TINY ORGANISMS that live just beneath the surface of LAKES, RIVERS, PONDS, and OCEANS across the planet.

2 The word PLANKTON comes from the Greek word for "DRIFTING".

This is because plankton can't swim against tides and currents, so they drift wherever they're taken.

3 There are two main types of plankton: PHYTOPLANKTON, which are PLANTS, and ZOOPLANKTON, which are ANIMALS.

They form the start of the watery food chain. Phytoplankton are eaten by zooplankton, which are in turn eaten by fish and some whales.

4 The individual organisms that make up plankton are called "PLANKTERS".

5 Some plankters are so TINY that a DROP OF SEAWATER may hold 100,000 of them.

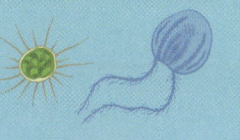

21 December
MODERN SURGERY

1 ADVANCES in medicine mean surgery is SAFER THAN EVER.
Surgeons have robots, microscopes, and lasers to help them with operations.

2 The FIRST successful ORGAN TRANSPLANT was a kidney (between twin brothers), in 1954.
Now hundreds of thousands of transplants are performed worldwide every year.

A surgeon can use a computer to control robotic arms.

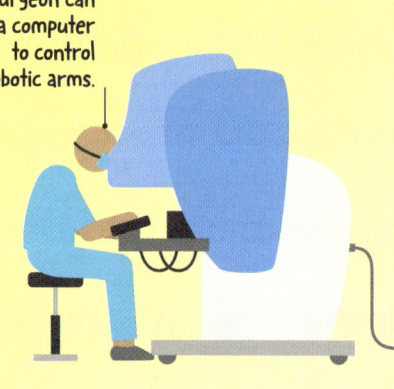

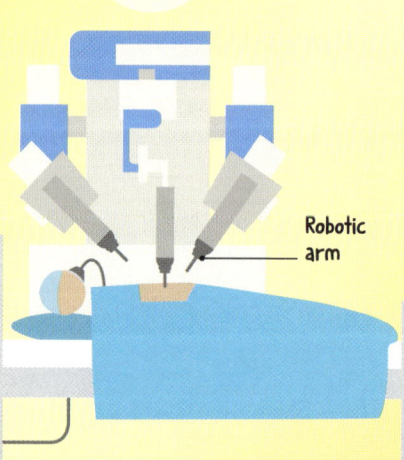

Robotic arm

3 Doctors can perform "TELESURGERY" from THOUSANDS OF MILES AWAY.
They use the internet and robots to carry out the operation, with no need to be in the room.

4 Surgeons first operated on TINY BABIES still in the MOTHER'S WOMB in the 1980s.

5 Surgeons have found ways to operate without making BIG INCISIONS (cuts).
Some surgeries can be performed using tiny tubes called catheters threaded through blood vessels!

---

1 Infrared radiation is a type of INVISIBLE LIGHT.
It is part of the electromagnetic spectrum (see p204).

2 Humans CAN'T SEE infrared waves, but we CAN FEEL THEM.
We feel infrared radiation as heat from the Sun or warm objects.

Infrared cameras can reveal the heat emitted by objects, such as this polar bear swimming in freezing water.

3 Some ANIMALS can SEE infrared light.
Snakes and mosquitoes use it to FIND FOOD in the dark.

4 Your TV remote control uses INFRARED WAVES to SEND SIGNALS to the TV.

Infrared images reveal the surface of Saturn's moon, Titan, beneath its thick atmosphere.

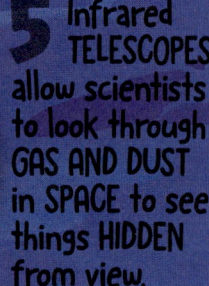

5 Infrared TELESCOPES allow scientists to look through GAS AND DUST in SPACE to see things HIDDEN from view.

22 December
INFRARED RADIATION

1

Just as nerves (see p10) are the body's high-speed messengers, HORMONES are its SLOWER, CHEMICAL MESSENGERS.

The hypothalamus releases hormones that stimulate the pituitary gland.

The pineal gland releases melatonin, which regulates your body clock.

2

There are more than 50 KNOWN HORMONES in the human body. There may be more yet to be discovered by scientists.

The pituitary gland is involved in many important bodily functions, such as growth.

The thyroid gland controls how quickly the cells in your body work.

The thymus gland controls part of the body's immune system.

The heart releases hormones that can help lower the body's blood pressure.

3

Special organs in the body called GLANDS RELEASE HORMONES into the blood.

Together, the glands and hormones are known as the endocrine system.

The stomach releases hormones to tell you that you're hungry.

The adrenal glands release hormones that help you respond to stress and danger.

The pancreas controls your blood-sugar levels.

Sex organs (ovaries in women and testes in men) release hormones important in reproduction (see p177).

23

December

HORMONES

4

Hormones REGULATE many BODILY FUNCTIONS.

These range from growth and reproduction to sleep patterns and whether you feel hungry or full.

5

Oxytocin is sometimes called the LOVE HORMONE.

It is released by cells in the brain and encourages loving feelings between people, including parents and their babies.

TELESCOPES

1

TELESCOPES magnify distant objects by GATHERING AND FOCUSING LIGHT.

Most non-professional ones use two main types of lens.

2

Large telescopes found in OBSERVATORIES use MIRRORS to focus the light.

They are located in remote areas where the air is still and the skies are very dark and clear.

OBSERVATORY TELESCOPE

5

Some telescopes collect WAVELENGTHS OTHER THAN VISIBLE LIGHT.

Radio telescopes use dish antennas to collect radio waves from distant stars and galaxies.

COMPACT ARRAY TELESCOPE

4

The ELT will detect 100 MILLION TIMES more light than a HUMAN EYE.

Scientists hope it will find Earth-like planets orbiting other stars.

3

The EXTREMELY LARGE TELESCOPE (ELT), under construction in the Atacama Desert, Chile, will be the biggest optical telescope in the world.

Its giant primary mirror will be 39 m (128 ft) wide.

REFRACTOR TELESCOPE

1. Light from a star enters the telescope.

2. A convex lens refracts (bends) the light.

3. A second convex lens focuses the image onto the eyepiece.

REFLECTOR TELESCOPE

1. Light enters the telescope.

4. A convex lens focuses and magnifies the image.

3. A second mirror reflects the light towards the eyepiece.

2. A concave mirror focuses and reflects the light back up the telescope tube.

258

1

Chocolate is made from BITTER COCOA BEANS, which have to go through MANY PROCESSES to make it SWEET AND DELICIOUS.

1. Chocolate is made from cacao beans from the fruit of the cacao tree.

2. The beans are fermented for several days to soften the pulp and develop the flavour.

3. The beans are dried in the Sun or a kiln to prepare them for shipping.

25
December
SWEETS

2

SUGAR and WATER are the BASIS of many different sweets.

Whether they are hard or soft depends on how the mixture is cooled.

4. Dried beans are roasted to improve their flavour, then the shells are removed.

5. The shelled beans (nibs) are ground into a paste, called chocolate liquor.

3

The ANCIENT EGYPTIANS ate marshmallows 4,000 YEARS AGO AS A MEDICINE, making this from the mallow plant.

5

There is a SCIENTIFIC REASON why you might CRAVE SUGAR.

Eating sugary foods releases chemicals that make us feel good, such as the hormone (see p257) dopamine.

8. The chocolate is poured into moulds and cooled, ready for eating.

7. Tempering (heating and cooling) ensures crystals in the chocolate are evenly sized so it looks smooth and shiny.

6. The chocolate liquor is mixed with ingredients such as sugar and cocoa butter.

4

POPPING CANDY BURSTS in your mouth because of TINY BUBBLES of carbon dioxide gas.

As the candy melts in your mouth, the gas is released!

1 There are about 18,500 KNOWN BUTTERFLY SPECIES.

2 Butterflies TASTE WITH THEIR FEET, using special sensors!

3 The WESTERN PYGMY BLUE is the SMALLEST BUTTERFLY IN THE WORLD, with a wingspan of just 1.3 cm (0.5 in).

The glasswing butterfly's transparent wings help it blend into its rainforest habitat.

4 The wings of the GLASSWING BUTTERFLY are almost COMPLETELY TRANSPARENT.

5 North American black and orange MONARCH BUTTERFLIES can FLY MORE THAN 4,000 km (2,500 miles) when they MIGRATE south during winter.

26
December
BUTTERFLIES

Vibrant colours and patterns are made up of tiny scales.

27
December
MOTHS

2 There are about 160,000 KNOWN MOTH SPECIES worldwide.

3 The ATLAS MOTH, of Southeast Asia, is the LARGEST in the world, with a wingspan of nearly 30 cm (12 in).

4 Not all moths are NOCTURNAL. The HUMMINGBIRD HAWK-MOTH is a day-flying moth that SIPS NECTAR from flowers.

5 The BIRD-DROPPING MOTH has evolved to LOOK LIKE BIRD POO to avoid predators.

The hummingbird hawk-moth uses a long proboscis (sucking mouthpart) to drink nectar.

1 MOTHS were around BEFORE BUTTERFLIES.

In fact, the first butterflies evolved from moths about 100 million years ago.

28
December
FORCES

1 FORCES are PUSHES and PULLS. They can make an object move or change speed, direction, or shape.

Gravity pushes the kayak down, but the water also pushes it up.

The paddle pushes against the water.

2

Forces come in PAIRS.

Every force has an equal and opposite force. Even if you stand still, there are two forces at work: gravity pulls you down, while the atoms and molecules in the ground resist being compressed, and push back.

The water pushes against the kayak, creating friction or drag.

The push of the paddle is stronger than the drag of the water, so the kayak moves forward.

5

ENGINEERS design things with FORCES IN MIND.

Racing car engineers design cars with streamlined shapes to reduce air resistance, while bridge engineers carefully balance a bridge's strength with the weight of its load, so it doesn't collapse.

4

Forces are INVISIBLE.

You might not be able to see them, but you can observe the effects they have on the things they act upon.

The Moon has just one-sixth of Earth's gravity, so you would weigh 16.6 per cent of your Earth weight there.

3

WEIGHT measures how much GRAVITY pulls on something.

29
December
DIGESTIVE SYSTEM

1 IT TAKES 24–72 HOURS for your digestive system to process food, FROM THE FIRST BITE TO EXPELLING THE WASTE.

Food is broken down in stages into nutrients that feed your body's cells.

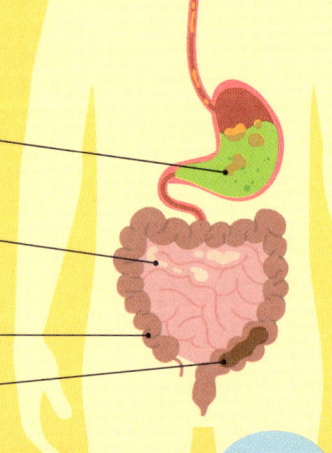

Chewing and swallowing takes up to a minute.

Breaking down food in the stomach takes 1–4 hours.

Passing through the small intestine takes 4 hours.

Passing through the large intestine takes 16+ hours.

Waste leaving the body takes about a minute.

2 Your whole digestive system, if stretched out, would be about 8 m (26 ft) long – THAT'S FOUR TIMES TALLER THAN AN AVERAGE HUMAN!

The digestive system runs from the mouth to the anus, with lots of interconnected tubes and chambers in between.

3 Digestion uses CHEMICALS CALLED ENZYMES to break food down into SMALLER, SIMPLER NUTRIENTS.

These include glucose (sugar) for energy; amino acids for growth; and fatty acids for repair. The nutrients travel in the blood to the body's cells.

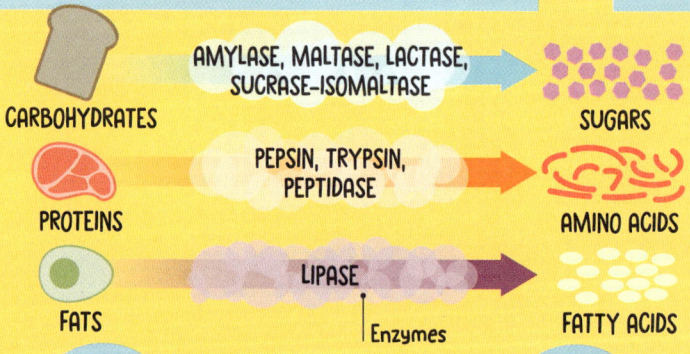

CARBOHYDRATES — AMYLASE, MALTASE, LACTASE, SUCRASE-ISOMALTASE → SUGARS

PROTEINS — PEPSIN, TRYPSIN, PEPTIDASE → AMINO ACIDS

FATS — LIPASE → FATTY ACIDS

Enzymes

4 Your digestive system will keep food moving EVEN IF YOU STAND ON YOUR HEAD!

This is because muscles in the walls of digestive organs automatically contract and relax in a wave-like rhythm to push food along, in a process called peristalsis.

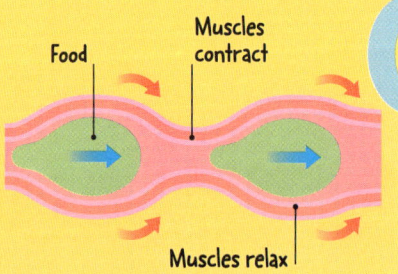

Food

Muscles contract

Muscles relax

5 NEWBORN BABIES can only digest one thing – MILK.

They can't chew or swallow, and their digestive systems have not yet developed enough to process anything other than milk.

30 December
GLASS

1 Most GLASS IS MADE BY MELTING A TYPE OF SAND CALLED SILICA with small amounts of soda ash and limestone, so it becomes transparent.

2 IF LIGHTNING HITS SAND, THE INTENSE HEAT CAN MELT THE GRAINS and fuse them together, CREATING A NATURAL FORM OF GLASS CALLED FULGURITE.

3 Natural glass can also be formed by METEORS STRIKING sand or rock, or when VOLCANIC LAVA COOLS FAST to form a BLACK GLASS called OBSIDIAN.

4 THE WORLD'S LARGEST GLASS WINDOW IS THREE GIRAFFES TALL.

Around 16.5 m (54 ft) high and 3 m (10 ft) wide, the window sits in the Taikang Tower in Beijing, China.

5 THE OLDEST HUMAN-MADE GLASS IS 4,000 YEARS OLD.

Humans first started making glass in ancient Mesopotamia. The earliest glass items were decorative beads, as well as bottles for precious oils and perfumes.

31 December
FIREWORKS

1 FIREWORKS are bundles of chemicals and GUNPOWDER.

The chemicals, or "stars", inside the firework are carefully arranged to create colourful patterns.

2 The GUNPOWDER in fireworks is called "BLACK POWDER".

It is made of charcoal, potassium nitrate (saltpetre), and sulfur.

3 Different MINERALS create the colourful BANGS AND SPARKLES.

For example, sodium is used for yellow, copper for blue, and barium for green.

4 The FIRST FIREWORKS were invented in CHINA, AROUND 2,000 YEARS AGO.

5 Fireworks have A QUICK FUSE AND A SLOW FUSE.

The quick fuse ignites the lift charge, to blast the rocket into the sky. The slow fuse ignites the "stars", creating a colourful display.

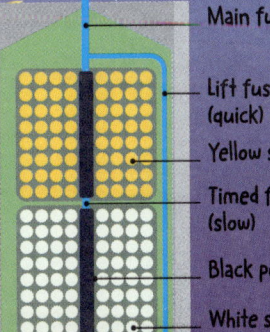

Red sparks are created by strontium.

- Main fuse
- Lift fuse (quick)
- Yellow stars
- Timed fuse (slow)
- Black powder
- White stars
- Lift charge

GLOSSARY

Acceleration
An increase or decrease in an object's speed due to a force being applied to it.

Acid
A reactive chemical that has a pH of less than 7 (see **pH**). Strong acids can cause harmful burns when they come into contact with humans, whereas weak acids are often found in ingredients such as vinegar and lemon juice.

Adrenaline
A hormone that prepares the body for sudden action at times of danger and excitement.

Air resistance
The force that pushes against objects as they move through the air; also known as drag.

Algae
Plant-like organisms that can make energy from sunlight and which mainly live in an aquatic environment.

Altitude
The height of an object above sea level.

Antennae
Long sensory organs on the heads of certain animals, such as insects and crustaceans.

Antibody
A substance that sticks to germs and marks them for destruction by white blood cells.

Artery
A blood vessel that carries blood away from the heart and to the body's tissues and organs, such as the liver.

Artificial intelligence (AI)
Computer systems designed to think and learn in order to perform tasks typically requiring human intelligence. AI is also the branch of computer science that develops these systems.

Asteroid
An small, irregular Solar System object made of rock and/or metal that orbits the Sun.

Atmosphere
The layer of breathable gases, such as oxygen and nitrogen, that surrounds Earth. Where the atmosphere ends, space begins.

Atom
The smallest unit of an element. The number of protons in an atom determine which element it is.

Bacteria
Microscopic organisms with a simple single-celled form. There are billions of bacteria in our bodies and in the world around us – some good, some harmful.

Base
A reactive chemical that has a pH of more than 7 (see **pH**). Bases are the chemical opposite of acids. A base that will dissolve in water is called an alkali.

Battery
An energy-storing device that produces an electric current when connected to a circuit. Some batteries can be recharged.

Blood vessel
A tube that carries blood through the body. The main types of blood vessel are arteries, capillaries, and veins.

Bone
The strong, hard body part made mainly of calcium minerals. There are 206 bones in the human body.

Carbon dioxide
A colourless, odourless gas that exists naturally in the atmosphere and is also emitted when burning fossil fuels. An increasing amount of this gas in the atmosphere is causing global warming.

Cartilage
A tough, flexible type of connective tissue that helps support the body and covers the ends of bones and joints.

Cell
The basic unit from which all living organisms, including humans, are made. The parts within a cell are called organelles. Animal and plants cells are slightly different to each other.

Chemical
An element, or a substance made of more than one element. A chemical is pure or the same all the way through – it is not a mixture. Water, iron, and oxygen are chemicals.

Chemical reaction
A process in which substances are changed into new substances, as atoms are rearranged.

Chlorophyll
A substance found in plants and some other organisms that makes them green and is used in photosynthesis.

Circuit
A path that electricity flows around. All electronic devices have circuits in them.

Circuit board
A component found in electronic goods, such as laptops and mobile phones, that contains one or more electrical circuits.

Climate change
Long-term shifts in the usual weather patterns of Earth or a particular area. Often refers to the severe effects caused by human actions.

Compound
A chemical substance in which two or more elements have bonded together.

Condensation
The process that occurs when matter changes state from a gas into a liquid.

Constellation
A named area of the sky (defined by the International Astronomical Union). Many are based around distinctive patterns of stars.

Crust (Earth's)
The thin, solid outer layer of the planet. It is made up of two types: continental crust and oceanic crust.

Crystal
A solid substance with a highly ordered shape. Diamonds and salt grains are crystals.

Current
A flow of a substance. An electric current is a flow of electrons. An ocean current is a flow of water in the ocean, driven by the wind or other factors.

Cytoplasm
A jelly-like substance that fills up a cell.

Data
Information that can be analysed, often in the form of facts or statistics. In computing, information that can be processed by a computer.

Decomposer
Bacteria, fungi, or other organisms that break down dead organisms in a process called decomposition, or rotting.

Density
The mass of a solid, liquid, or gas per unit of volume. A dense material has lots of atoms packed closely together. Less dense objects float in more dense fluids, such as how wood can float in water because it is less dense than water.

Disease
Any problem with the body that makes someone unwell. Infectious diseases are those caused by germs.

DNA
Short for "deoxyribonucleic acid", a long, thin, double-helix shaped molecule found in the cells of all living organisms. It carries genetic code, the instructions for how a living thing will look and function.

Drag
See **Air resistance**.

Ecosystem
A community of living organisms that interact with each other and their environment.

Egg
The female reproductive cell or a hard-shelled body laid by animals such as birds.

Electricity
Anything related to electric charge. Electric current is the movement of particles with electric charge. Electricity is used to power homes, cars, and many other modern machines.

Electromagnetic spectrum
The whole range of different types of electromagnetic radiation, from gammas rays to radio waves. Visible light, which we can see, is part of the electromagnetic spectrum.

Electron
A negatively charged particle found in the outer part of an atom. Moving electrons carry electricity and cause magnetism.

Element
A substance made of only one kind of atom. There are 118 known elements, about 90 of which occur naturally.

Embryo
The term used to describe a developing baby in the first eight weeks following fertilization.

Energy
What enables work to be done. Energy exists in many different forms, such as electrical and chemical. It cannot be created or destroyed, only transferred.

Equator
An imaginary circle around the centre of Earth, dividing it into the northern and southern hemispheres (halves).

Evaporation
A process by which a liquid changes into a gas.

Evolution
The gradual process of change in living things between generations over millions of years.

Fertilization
The joining of male and female sex cells to produce a new living thing as part of reproduction.

Fertilizer
A natural or artificial substance that is put on land to make plants grow better.

Fluid
A substance that can flow, such as a gas or liquid.

Force
A push or pull that causes an object to change its speed, direction, or shape.

Fossil
The preserved remains or traces of animals or plants from an earlier time. Fossils form over millions of years and can come in many different forms.

Fossil fuel
A fuel derived from the fossilized remains of living things. Fossil fuels include coal, oil, and gas. They release carbon dioxide and other harmful gases when they are burned, contributing to global warming.

Frequency
A measure of waves – the number of waves that pass a point every second.

Friction
The dragging force that occurs when one object moves over another. Friction slows down moving things.

Fruit
The ripened ovary of a flower, containing one or more seeds. Some fruits are sweet and juicy to attract animals.

Fuel
A substance used to provide heat or power to a machine or organism.

Fungi
A group of microorganisms that break down plants and animals. Some fungi have medicinal uses. Mushrooms and toadstools are fungi.

Galaxy
A collection of millions or trillions of stars, gas, and dust held together by gravity.

Gas
A state of matter where the particles are spaced out and move around at high speed. Gases flow to fill a container and can be compressed.

Gears
Mechnical devices such as interlocking cogs that make the turning effect of a force bigger or smaller.

Gene
A length of DNA that carries code to perform a specific job. Genes instruct cells to make proteins that affect one or more of the organism's characteristics. Genes are passed down from one generation to the next.

Germ
A tiny living thing (microorganism) that can get into the body and cause illness. Bacteria and viruses are types of germ.

265

Global warming
A rise in the average temperature of Earth's atmosphere, caused by rising levels of carbon dioxide and other greenhouse gases. Global warming has many effects on Earth's climate, including the melting of ice at the poles, a rise in sea levels, and the occurence of more extreme weather events.

Gravity
A force that pulls all things with mass towards each other. On Earth, gravity pulls objects to the ground and gives them weight. The planets of the Solar System are kept in orbit by gravity.

Greenhouse effect
A process where gases in the atmosphere trap heat from the Sun around the planet, warming up Earth.

Habitat
The natural home of an animal or plant.

Hormone
A chemical produced in the human body to change the way part of the body works. Hormones are carried by the blood.

Hurricane
A type of violent storm also known as a tropical cyclone.

Immune system
The defence mechanism of the human body that protects us from diseases by searching out and destroying threats.

Invertebrate
A type of animal that does not have a backbone.

Joint
A connection between two bones. Move joints, such as the hip and shoulder joints, allow the body to move around.

Lava
Hot, molten rock erupted from volcanic vents or fissures.

Lens
An object with curved surfaces that causes light to bend as it passes through. Lenses are used to form images in cameras, telescopes, and microscopes.

Magma
Hot, liquid rock that is found beneath Earth's surface.

Magnet
An object that has a magnetic field and attracts or repels other magnetic objects. Things are attracted or repelled by magnets due to an invisible force called magnetism.

Magnetic field
The invisible pattern of force that stretches out around a magnet. Earth is surrounded by a magnetic field.

Mantle
The semi-molten layer of Earth that sits below the crust.

Mass
A measure of the amount of matter in an object.

Matter
The material which everything is made from. Anything that has mass and occupies space is matter. Matter includes solids, liquids, and gases, and both living and non-living things.

Microorganism
A tiny organism that can be seen only with the aid of a microscope.

Mineral
Any naturally occuring substance made up of more than one element that can be found in the ground.

Mixture
A collection of substances that fill the same space but are not connected by chemical bonds.

Molecule
A group of two or more atoms joined by strong bonds.

Motor
A device used to generate movement, often powered by electricity.

Muscle
An organ in the body made up of tiny fibres. Muscles pull on bones to make the body move.

Nectar
A sugary liquid produced by flowers. Bees collect nectar to make honey.

Nerves
A bundle of specialized cells that carry electrical signals around an organism's body.

Neuron
A term for a nerve cell. Neurons carry information around the body as electrical signals.

Nuclear fusion
A process in which small atomic nuclei, such as those of hydrogen atoms, join together to make larger ones, releasing large amounts of energy.

Nucleus
The central core of something. The nucleus at the centre of an atom contains protons and neutrons, while a cell's nucleus contains DNA.

Nutrient
A substance that animals and plants take in that is essential for life and growth. Nutrients are useful as a source of energy or a raw material.

Orbit
The path taken by an object – for example, a planet – that is moving around another object.

Organ
A major structure in an organism that has a specific function. Examples include the stomach, the brain, and the skin.

Organism
Any living thing, such as a plant, animal, or fungus.

Oxygen
A colourless, odourless gas found in Earth's atmosphere. It is required for a substance to burn.

Parasite
An organism that feeds on another, called the host, weakening it, and sometimes eventually killing it.

Particle
A tiny speck of matter.

Pathogen
A microorganism, such as a virus, that causes disease. Pathogens usually reproduce inside a host and then infect others.

pH
A scale used to measure how acidic or alkaline a solution is.

Photosynthesis
The process by which plants use the Sun's energy and chlorophyll to make food molecules from water and carbon dioxide. It takes place within plant cells and produces oxygen and sugars.

Plankton
Organisms that drift in the ocean, rather than swimming against the current. Most plankton are tiny, but they exist in large numbers. They are an important source of food for bigger animals.

Plastic
A type of polymer that has a wide range of useful properties.

Pollen
Tiny grains produced by flowers, which contain the male cells needed to fertilize eggs. Insects unwittingly carry pollen from one flower to another, helping plants reproduce.

Pollination
The transfer of pollen from the male part of a flower to the female part of a flower. Pollination is essential for sexual reproduction in flowering plants.

Polymer
A long, chain-like molecule made of smaller molecules connected together. Polymers can be found in nature, such as DNA, or be produced artificially.

Precipitation
Water that falls from the clouds to the ground, such as rain, snow, hail, and sleet.

Predator
An animal that hunts other animals for food.

Pressure
The force exerted by something pressing or squeezing an area. The same force can produce high pressure or low pressure depending on the area it acts on.

Prey
An animal hunted by other animals for food.

Protein
A type of complex chemical found in all living things. Proteins are the building blocks of cells, and organisms need them for growth and repair. Foods such as fish, cheese, and beans contain protein.

Radiation
Waves of energy that travel through space. Radiation includes visible light, heat, X-rays, and radio waves.

Radioactive
A material that is unstable because the nuclei of its atoms easily break down.

Radio wave
A type of electromagnetic radiation that has the longest waves. It travels far and very quickly, and can be used to carry information, such as music.

Renewable energy
A type of energy that comes from a source that will not run out, unlike energy from fossil fuels. Types of renewable energy include wind power, wave power, and solar power.

Reproduction
The process by which plants and animals produce new offspring.

RNA
Ribonucleic acid, a molecule similar to DNA. Some viruses have RNA instead of DNA.

Room temperature
A standard scientific term for comfortable conditions (for humans), usually a temperature of around 20°C (68°F).

Satellite
An object in space that travels around another in an orbit. Many satellites are human-made.

Sediment
Small particles of rock that are deposited by water.

Seed
A reproductive structure containing a plant embryo and a food store.

Solar System
The Sun and all the bodies held in orbit around it by the Sun's gravity. These bodies include planets, dwarf planets, moons, asteroids, and comets.

Streamlined
Smoothly shaped to move easily through air or water.

Sugar
A usually sweet-tasting substance needed by cells to live and grow.

Synthetic
A human-made chemical.

Tectonic plates
Giant pieces of Earth's crust, which move around over millions of years.

Temperature
A scientific measure of how hot or cold something is.

Tissue
A group of similar cells that carry out the same function, such as muscle tissue, which can contract. Many tissues make an organ.

Toxin
A poisonous substance.

Turbine
A device with rotating fan blades that are driven by the pressure of gases, liquids, or steam. It converts energy into a different form.

Ultrasound
Sound with a frequency above that which the human ear can detect.

Ultraviolet
A type of electromagnetic radiation with a wavelength shorter than visible light.

Universe
The whole of space and everything it contains.

Upthrust
The upwards force exerted by a liquid or a gas on an object immersed in it.

Vaccine
A substance usually containing weakened or dead pathogens that stimulates the production of antibodies in a person's body.

Vacuum
A space in which there is no matter.

Vein
A blood vessel that carries blood towards the heart.

Velocity
A measure of an object's speed and direction.

Vertebrate
An animal with a backbone.

Virus
A tiny, infectious non-living agent that causes disease by invading and multiplying inside body cells.

Volume
The amount of space an object takes up.

Wave
Vibration that transfers energy from place to place, without transferring the matter that it is flowing through.

Wavelength
The distance between two successive peaks or two successive troughs in a wave.

Weight
The force applied to a mass by gravity.

White blood cell
A cell found in the blood that is involved in defending the body against pathogens.

Index

Acknowledgements

DK would like to thank the following for their help with this book:
Catharine Robertson and Carron Brown for proofreading; Elizabeth Wise
for compiling the index; and Mik Gates for additional design assistance.

Picture Credits

The publisher would like to thank the following for their kind permission to reproduce their photographs:

(Key: a-above; b-below/bottom; c-centre; f-far; l-left; r-right; t-top)

3 Shutterstock.com: Incomible (bl). 5 Shutterstock.com: Incomible (bc). 8 ESA / Hubble: Davide De Martin & the ESA / ESO / NASA Photoshop FITS Liberator (ca); NASA, ESA, and the Hubble Heritage Team (AURA / STScI) (cb). NASA: ESA, M. Robberto (Space Telescope Science Institute / ESA) and the Hubble Space Telescope Orion Treasury Project Team (tl); ESA and the Hubble Heritage (STScI / AURA)-ESA / Hubble Collaboration (br). 9 Getty Images / iStock: Antagain (clb); joegolby (bl). 11 Dorling Kindersley: Phil Gamble. Dreamstime.com: Ken Backer (ca/x2). 13 Dreamstime.com: Jessica Evans (cra). 15 Dreamstime.com: Robin Winkelman (cra). 17 Dreamstime.com: Andamanse (tr). Shutterstock.com: muratart (t). 19 Dreamstime.com: Dalius Baranauskas (bl). 22 123RF.com: Sergey Goruppa (bc). Dorling Kindersley: Adrian Whicher / Science Museum, London (tr). 24 Dreamstime.com: Brett Critchley (bl). ESA / Hubble: NASA, ESA and Allison Loll / Jeff Hester (Arizona State University) / Davide De Martin (t). 25 123RF.com: phive2015 (t). 28 NASA: ESA, CSA, STScI (bl). 31 Dreamstime.com: Releon8211 (l); Stockshooter (bl). NASA: Johns Hopkins University Applied Physics Laboratory / Carnegie Institution of Washington (cr, crb). 32 Alamy Stock Photo: imageBROKER.com GmbH & Co. KG / Richard Becker (br). 34 Dreamstime.com: Anetlanda (bc). 35 Dreamstime.com: Mathieu Le Mauff (bl). Unsplash: Gordon Mak (cla). 36 Dorling Kindersley: Liberty's Owl, Raptor and Reptile Centre, Hampshire, UK (crb/tarantula); Linda Pitkin (crb). Dreamstime.com: Eastmanphoto (cl); Isselee (cla); Vasyl Helevachuk (ca); Usensam2007 / Roman Samokhin (ca/gorilla). Getty Images / iStock: EXTREME-PHOTOGRAPHER (c). 39 Getty Images / iStock: Divelvanov (bl). 40 NASA: CXC / MIT / L.Lopez et al.; Palomar Observatory / Caltech; NSF / NRAO / VLA (br). Shutterstock.com: RugliG (l). 41 Dreamstime.com: Nilofers (cb). 43 Fotolia: apttone (crb). Getty Images: George Pachantouris (bl). 45 Dreamstime.com: Antartis (l); Chuyu (bc). 46 Dreamstime.com: Martin Holverda (br). 47 Alamy Stock Photo: Martin Harvey (bc). Dreamstime.com: Annaav (c); Jlvdream / Joao Virissimo (clb); Nattle (tc). 48 Dreamstime.com: Derek Gordon (crb). 49 Dreamstime.com: Appfind (l); Eris Isselee (cb). Getty Images / iStock: powerofforever (cra). 50 Dreamstime.com: Andrei Nekrassov - anekrassov@gmail.com (bc). 51 Alamy Stock Photo: Science Photo Library / Steve Gschmeissner (cra). Dreamstime.com: Rkpimages (tl). Fotolia: Olena Pantiukh (clb). Getty Images / iStock: Andyworks (bc). 52 Dorling Kindersley: Andy and Gill Swash (cb). Dreamstime.com: Donyanedomam (crb). 55 Dreamstime.com: Ebastard129 (cra). 56 Dreamstime.com: Nui7711 (crb). NASA: NASA Visualization Technology Applications and Development (VTAD) (c). 57 123RF.com: teerayut ninsiri (br); peterwaters (x8). 58 Dreamstime.com: Bazruh (cb). 59 Alamy Stock Photo: imageBROKER.com GmbH & Co. KG / Thomas Dressler (t). Dreamstime.com: Snake3d (cb). 61 123RF.com: Oleksiy (tl, cl, bl); Yuliia Pushkar (l). Dreamstime.com: Jgade (cr). Photolibrary: White / Digital Zoo (br). 62 Dreamstime.com: Stnazkul (bl). 63 Dreamstime.com: Akesin (cb). 64 Getty Images / iStock: Philip Stewart (cb); tonaquatic (clb). 65 Dreamstime.com: Isselee (cl, b/x4); Anthony Aneese Totah Jr (br); Dmytro Skrypnykov (tr). 67 123RF.com: Petri Jauhiainen. NASA and The Hubble Heritage Team (AURA/STScI): ESA, and A. Schaller (for STScI) (br). 68 123RF.com: Eric Isselee (cla). 69 123RF.com: Monika Wisniewska (cla). Getty Images / iStock: anankkml (tr). 70 Dreamstime.com: Colette6 (c). 71 Science Photo Library: MSF / Javier Trueba (cr). Shutterstock.com: Incomible (crb). 73 Shutterstock.com: Nataly Studio (br). 76 Dreamstime.com: Dibrova (br); Tebnad. 77 123RF.com: yelo34 (tc). Dreamstime.com: Artushfoto (bl). 78 Dreamstime.com: Kamnuan Suthongsa (ca). Getty Images / iStock: GlobalP (cr); pchoui (tr). Shutterstock.com: sciencepics (cla). 79 Dreamstime.com: Philip Openshaw (bl). 82 123RF.com: leonello calvetti (ca). 83 Dreamstime.com: Yulia Kuleshova (b); Smgirly / Simone Gatterwe (tc); mhprice / Mike Price (tr). Shutterstock.com: Shaun Jeffers (clb). 84 Alamy Stock Photo: Nature Picture Library / Alex Mustard (bc). 85 NASA: (tr). 87 Dreamstime.com: Sarah2 (tc). 89 Dreamstime.com: James Billimore (cb). 90 123RF.com: joytasa. 91 Dorling Kindersley: Dan Crisp (t/x14). Shutterstock.com: Liang Li Photos (cra). 93 Dreamstime.com: Joris Van Den Heuvel (br); David Maixner (t); Natalyka (cra); Aleksandr Kurganov (b). 94 Getty Images / iStock: E+ / Just_Super (b). 96 Dreamstime.com: Seadam (t). 97 NASA. 98 123RF.com: Duncan Noakes (bl). Dorling Kindersley: Booth Museum of Natural History, Brighton / Dave King (cra). Fotolia: Mark Higgins (ca). 100 Alamy Stock Photo: MR3D (tr). NASA: Johns Hopkins University Applied Physics Laboratory / Southwest Research Institute (bl); JPL-Caltech / UCLA / MPS / DLR / IDA (c). Science Photo Library: Mark Garlick (cra, clb). 101 123RF.com: pinipin / Phil Gamble (cra, cb, crb). 102 Getty Images / iStock: E+ / bjdlzx. 104 Dorling Kindersley: Blackpool Zoo, Lancashire, UK (fcrb). Dreamstime.com: Ded350let (bl); Isselee (crb); Ondrej Prosicky (tr). Fotolia: Shchipkova Elena (cl). 105 Dreamstime.com: Mikhailsh (ca). 106 Dreamstime.com: Alicenerr (br). 107 Shutterstock.com: Schager (clb). 109 Dreamstime.com: Manav Lohia (tr). Getty Images / iStock: Jennifer_Sharp (bl). 111 Dreamstime.com: Ralf Lehmann (ca). NASA: JPL-Caltech (br). Shutterstock.com: Siwakorn1933 (t). 112 Dreamstime.com: Alejandro Miranda (cra). NASA: GSFC / Arizona State University (ca). 113 Getty Images / iStock: ratpack223 (br). 114 Shutterstock.com: Puwadol Jaturawutthichai (cla). 117 123RF.com: photodeti / Ermolaev Alexandr Alexandrovich (tl). Dreamstime.com: Isselee (crb); Sanchai Rattakunchorn (b). 121 Dreamstime.com: Mr.smith Chetanachan (c). 123 123RF.com: andreahast (cra). 124 Dreamstime.com: Anteroxx (cla). 125 Getty Images / iStock: alexey_boldin (bc). Science Photo Library: Steve Gschmeissner (r). 126 123RF.com: Brian Kinney (clb). 127 Dreamstime.com: Dmitry Naumov (tc). 130 Alamy Stock Photo: Adisha Pramod (crb). Dreamstime.com: Rudmer Zwerver (tr). 131 Dorling Kindersley: EMU Unit of the Natural History Museum, London / Sue Barnes (bl). 132 Dreamstime.com: Vjanez (x4, br). NASA: Enhanced image by Kevin M. Gill (CC-BY) based on images provided courtesy of NASA / JPL-Caltech / SwRI / MSSS (ca). 133 Shutterstock.com: BlueRingMedia (bc). 136 Dreamstime.com: Edgloris Marys (cra). 137 Shutterstock.com: Dan-Pepper (bl); Daniel Vidal T (ca); ThomasChoi (b). 138 Shutterstock.com: Benny Marty (br). 139 123RF.com: jovannig (tc). Alamy Stock Photo: UPI (bc). Shutterstock.com: Peakstock (tl). 140 Dreamstime.com: Jpsdk / Jens Stolt (tl); Luis Leamus (cb); Welshi23 (clb). 141 123RF.com: 2nix / Panithan Fakseemuang (br). Dreamstime.com: Kirati Kicharearn (r/x11). 142 123RF.com: Michael Rosskothen (crb). Dreamstime.com: Mark Turner (cra). 143 Shutterstock.com: ZinaidaSopina (crb). 144 Science Photo Library: Mark Garlick (cra). 145 123RF.com: Marina Scurupii (r). Pixabay: Camera-man / Sergio Cerrato - Italia (bc). 146 Dorling Kindersley: ESA / James Stevenson (cla). 147 123RF.com: smileus (br). Dorling Kindersley: Royal Pavilion & Museums, Brighton & Hove (bc). 149 NASA: (tl, bc). 150 Dreamstime.com: Eduard Muzhevskyi (br). 151 Dreamstime.com: Rafael Ben Ari (bl); Aliaksandr Mazurkevich (fcl). Getty Images / iStock: YinYang (br). 152 Dorling Kindersley: Harry Taylor / Natural History Museum, London (clb). Dreamstime.com: Joools (cl). 155 123RF.com: alekss / Alexandr Pakhnyushchyy (cl). Dreamstime.com: Roiandroi (bl). NASA and The Hubble Heritage Team (AURA/STScI): NASA, ESA, N. Smith (University of California, Berkeley), and The Hubble Heritage Team (STScI / AURA); (r). NASA: ESA, M. Robberto (Space Telescope Science Institute / ESA) and the Hubble Space Telescope Orion Treasury Project Team (bc). 157 Dreamstime.com: Nexusplexus (b); Damrong Rattanapong (ca). 158 123RF.com: lightwise (t). Alamy Stock Photo: Johan Möllerberg (bl). Dreamstime.com: Eddydegroot (b). 159 Dorling Kindersley: The Flag Institute / Simon Mumford (clb). Getty Images / iStock: MindStorm-inc (crb). 160 Alamy Stock Photo: RGB Ventures / SuperStock / Mark Newman (cb). Dorling Kindersley: Twan Leenders (cr). Dreamstime.com: Kotomiti_okuma (cra). 161 Dorling Kindersley: Dan Chambers / Gary Ombler (bc). 163 123RF.com: Daniel Prudek (cb/x3). Dorling Kindersley: Wildlife Heritage Foundation, Kent, UK (crb). Dreamstime.com: Beibaoke1 (br); Abeselom Zerit (bl); Rudolf Ernst (cla). 164 Dreamstime.com: Volodymyr Kucherenko (tr); Kayla Rowens (cra); Sander Meertins (clb). 165 Science Photo Library: HENNING DALHOFF (cra). 167 Alamy Stock Photo: Reinhard Dirscherl (tr); dotted zebra (b). Dorling Kindersley: Eleanor Bates (tc). 168 Dorling Kindersley: Robert Royse (br). Dreamstime.com: Mikelane45 (clb); Marcin Wojciechowski (cb). Getty Images: Daniel Parent (ca). 170-171 123RF.com: 1xpert / Phil Gamble (bc). 170 123RF.com: 1xpert / Phil Gamble (bl). Dreamstime.com: Uwe Bergwitz (l); Aleksej Sarifulin (cb). 172 Dreamstime.com: Sorin Colac (ca); Dmitryp / Dmitry Pichugin (cb). 173 Corbis: Warren Faidley (cla). Dreamstime.com: Yevhenii Tryfonov (t). 174 Dreamstime.com: Farek (br); Kkaplin / Kira Kaplinski (bl). 176 Dreamstime.com: Underworld (cr). 180 Dorling Kindersley: University of Aberdeen / Gary Ombler (bc). Dreamstime.com: Krzysztof Slusarczyk (t). Getty Images / iStock: 3DSculptor (b). 182 Dreamstime.com: Oleksandr Kalinichenko (ca); David Myslivec (cra). 183 Dreamstime.com: André Costa (bl); Orlando Florin Rosu (br); Chase Dekker (cla); Andrei Calangiu (ca). 184 Dreamstime.com: Nmcavaney (cla); Planetfelicity (clb); Christian Weiß (bc). 186 Dreamstime.com: Rixie (ca). Getty Images / iStock: Ultraforma (crb). 188 123RF.com: alekss / Alexandr Pakhnyushchyy (c). 189 123RF.com: Elena Duvernay (crb); Mark Turner (cra). 190 Dorling Kindersley: Natural History Museum, London / Harry Taylor (tc); Linda Pitkin (tl, ca). Dreamstime.com: Brad Calkins / Bradcalkins (cla); Izanbar (bc). 191 Shutterstock.com: F. Enot (b); PI (cla); VectorPlotnikoff (ca). 194 Dreamstime.com: Andreykuzmin (clb); Anikasalsera (br). Shutterstock.com: Spreadthesign (tr). 195 123RF.com: Wang Aizhong / Zhengzaishanchu (cb). Shutterstock.com: Alta Oosthuizen (bc). 196 Shutterstock.com: SurabhiArtss. 198 123RF.com: Wang Aizhong / Zhengzaishanchu (cb). Dorling Kindersley: Natural History Museum, London / Colin Keates (cl). Dreamstime.com: Chatcameraman (l). Getty Images: Photodisc / Eveleigh (ca). 200 Dreamstime.com: DreamStockIcons (tl). Getty Images: Pat Gaines (bc). 201 Dreamstime.com: Dr Ajay Kumar Singh (b). 203 Dreamstime.com: CreativeEndeavors (tr). 204 Dreamstime.com: Viesturs Davidčuks (clb). 205 Dreamstime.com: Dmstudio (tl). 207 Dreamstime.com: Taaeepang (cra). 209 Alamy Stock Photo: Chris Martin (cra). Shutterstock.com: Angel DiBilio (crb). 211 Alamy Stock Photo: Planetpix (tr). Dreamstime.com: Krzysztof Odziomek / Crisod (bc). 212 Dreamstime.com: Richard Carey (cla); Ljubisa Sujica (tr). 215 Dorling Kindersley: James Kuether (cl, br). Dreamstime.com: Marciomauro (cra). 216 Science Photo Library: STEVE GSCHMEISSNER / SCIENCE PHOTO LIBRARY (cra). 217 Dreamstime.com: Yinan Zhang / Cyoginan (ca). 220 Dorling Kindersley: RGB Research Limited / Ruth Jenkinson (clb). 221 Dreamstime.com: Xi Zhang (bl). Getty Images / iStock: Hakinci (cla). 224 Dorling Kindersley: Twan Leenders (cla). Dreamstime.com: Dudlajzov (bl); Sebastian Kaulitzki / Eraxion (ca); Kingmaphotos (cra). Fotolia: Steve Lovegrove (br). 225 123RF.com: Jakobradlgruber (cla). Alamy Stock Photo: Artem Bolshakov (c). Dreamstime.com: SaveJungle (bl). 226 Shutterstock.com: Prostock-studio (cla); VectorPlotnikoff (cb). 230 Dreamstime.com: William87 (fcla). Getty Images / iStock: Yuriy Kotsulym (fcl); Cheryl Ramalho (bl). 231 Shutterstock.com: GoodStudio (ftl); SpicyTruffel (ca). 236 123RF.com: Mark Grenier (c/background). 237 Dreamstime.com: Xunbin Pan / Defun (clb); Zakalinka (ca/background). Getty Images / iStock: johnandersonphoto (fcl). 242 Dreamstime.com: Gary Hanna / Tuktop (ftl); Isselee (cra); Jesse Kraft (fbl). Getty Images / iStock: DKart (cla). Shutterstock.com: Anna Kucherova (bc). 243 Dreamstime.com: Vladvitek (bl). Getty Images / iStock: Vasilina Shevchenko (tl). 244 Dreamstime.com: Mishoo (ca); Tatus (clb); Vitas (cb); Sabri Deniz Kizil / Bogalo (cr, cr/1). 245 Dreamstime.com: Irochka (c); Dmitry Mizintsev (c/background). 246 Dreamstime.com: Somphop Ruksutakarn (bl). 247 Dreamstime.com: Darren Baker (cb). ESA / Hubble: Davide De Martin (ca). 248 Dreamstime.com: Alslutsky (bl). 252 123RF.com: scanrail (crb). Dorling Kindersley: Ruth Jenkinson / RGB Research Limited (cra). Dreamstime.com: Valeria Head (bc); Retouch Man (tl). 253 Dreamstime.com: Sabri Deniz Kizil / Bogalo (cr, cr/1). Getty Images / iStock: Byrdyak (tl). 255 Dreamstime.com: Cat Vec (17/cr). 256 NASA/JPL/University of Arizona/University of Idaho: (br). Shutterstock.com: Maximillian cabinet (bl). 258 Dreamstime.com: Carol Buchanan / Cbpix (cr). Shutterstock.com: Veronique Duplain (fbl). 260 123RF.com: ankorlight (tc). Dreamstime.com: Janice Mccafferty / Mcjanice (cra); Vlasto Opatovsky (cb). 261 Shutterstock.com: ne2pi (c); Wth (bc)